A2 Chemistry
Max Parsonage

FACTS & PRACTICE FOR A2 LEVEL

FOR OCR

Maxinterax

Published by Maxinterax Ltd

12 Long Lane, Littlemore, Oxford OX4 3TW

© Max Parsonage 2009

The moral rights of the author have been asserted

First published 2009

All rights reserved. No part of this publication may be reproduced, stored in a retrieval system, or transmitted, in any form or by any means, without the prior permission in writing of Max Parsonage or as expressly permitted by law, or under terms agreed with the appropriate reprographics rights organization. Enquiries concerning reproduction outside the scope of the above should be sent to Max Parsonage, at the address above

You must not circulate this book in any other binding or cover and you must impose this same condition on any acquirer

British Library Cataloguing in Publication Data

Data available

ISBN 978-0-9555451-2-2

Edited and typset at Maxinterax Ltd

Printed in England

Author's Acknowledgements

I am pleased to recognise the special contributions of a few, amongst many others, that have made this book more accurate, clear, and accessible. My thanks go to my wife Jane for her editing, checking and encouragement, Dr. Steve Field for checking the chemical accuracy and question answers.

The cover graphics are used with the permisson of the artist Jane Parsonage who owns the rights to the artwork.

I am grateful for the comments and enthusiasm of many students.

INTRODUCTION

My aim in writing this book was to make chemistry both concise and clear, so that students can quickly reinforce class work, and subsequently revise it.

Through my role as an A Level Chemistry teacher I have been able to test successive drafts of this material with students, and this has been extremely valuable and interesting. It was a huge challenge to condense the essence of the subject into so few pages, but I know from working with students the value of doing this. There is just as much emphasis in the book on questions as there is on content, as only by testing themselves do students really discover how good their understanding of a topic is.

Exam-style questions are provided for students to practise their exam technique. A variety of styles is presented to reflect the diversity of questions in the papers. Students are given realistic space to write their answers, as in real exam questions. Students and teachers will be able to diagnose any exam technique problems by looking over the written answers.

This new, fully revised edition, uses more space to layout the topics as this was requested by readers.

MAX PARSONAGE

Max Parsonage is Head of Chemistry at d'Overbroeck's, an independent school in Oxford. He also writes about science for children, and develops interactive educational science software. He has recently helped produce educational software for the European Space Agency, and for Oxford University Press. In addition, he has authored and presented science vidoes.

CONTENTS

Why study chemistry?	vi
How to use this book	vii
How to succeed in your studies	vii
How to succeed in exams	viii
The A level system	ix
Exam specifications and modules	ix
The new A*grade	ix
1 Arenes	2
2 Carbonyl compounds	8
3 Carboxylic Acids, Esters and Amines	14
4 Polymers, optical isomers, proteins and amino acids	20
5 Synthesis	26
6 Spectroscopy and Chromatography	32
7 How fast? Rates	38
8 How far? Equilibrium calculations	44
9 Equilibrium: acids, bases and buffers	50
10 Lattice enthalpy	56
11 Enthalpy and entropy	62
12 Electrode potentials and fuel cells	68
13 Transition elements: theory	74
14 Transition elements: reactions	80
15 Synoptic: linking chemistry together	86
16 Experimental skills	92
Answers	98
Index	108
Periodic table	112

WHY STUDY CHEMISTRY?

Chemistry is fundamental to understanding the world around us, simply because everything is made of chemicals. From planets to cosmetics, and microbes to bridges, chemistry underpins how materials behave. It also explains how different substances can be made.

If you study A level chemistry then you should be able to ask 'Why?' and receive satisfying explanations. You will find AS chemistry explains chemical ideas mostly using words, while A2 chemistry explains chemical ideas using maths as well as words.

If you like logic problems, the way ideas can just 'click' beautifully together, then you will enjoy chemistry. Once you have gained a good grasp of the chemical patterns you will find there is very little detail to memorise, because studying chemistry is like studying a game. Once you know the 'rules of chemistry', you can 'play' with the chemical ideas. Chemistry is therefore a concise subject. It is attractive because it makes you think, without requiring you to write many essays or memorise huge amounts of information. Studying chemistry complements A levels that are essay based or require a huge reading load.

You may have to take chemistry if you want to become a doctor, or vet, or if you want to study chemistry, pharmacology, environmental science, or related subjects. Chemistry is also a useful foundation for biochemistry, geology, physical geography, engineering, or materials science at university.

Because it is such a fundamental study, chemistry provides helpful background for a great variety of subjects, such as biology, pyrotechnics in theatre studies, and art restoration. If law interests you, then chemistry is a useful discipline because it encourages logical thinking.

HOW TO USE THIS BOOK

This book will supplement a standard A level textbook, or you could use it as a free-standing A level book to consult alongside your notes. It is best used as a course companion, to be referred to when starting a topic, and to help you understand when you are in difficulties. When tests and exams approach it will usefully explain things in a few pages, and test you thoroughly.

Teachers may set the book as homework, or use it in class tests.

HOW TO SUCCEED IN YOUR STUDIES

To succeed in A level chemistry you, the student, need to **understand** the ideas, **remember** the facts and ideas, and have a good **exam technique**.

An **understanding** of chemistry builds up layer upon layer, so the units are laid out with the foundation topics first. For easy access, the units are presented in the order that you meet them in the spcifications. Units 1 to 4 (Organic chemistry) will help you understand units 5 and 6. Similarly the AS topics will help with the A2 ones when studying Lattice Enthalpy, Entropy, Equilibrium, and Rates. Understanding groups 2, and 7 will help you with Transition Metals.

Cover one topic at a time to gain a full understanding, rather than scanning many topics quickly.

To **understand** the topics, read the relevant section. If you do not immediately understand an explanation, pause and re-read it. Use the examples to help you see the point. Write short answers to the recall questions, then test yourself using the concept questions to check you understand the ideas, and then refer to the answers.

To **remember** the facts and ideas, use the factual recall questions (the 'Recall test'). If you can answer all the questions in a section, then you have the facts you need to answer the exam questions on that topic (the 'Concept test'). These questions are a more effective way of memorising information than simply copying notes. You could use them to help you to identify your weaknesses, then return to the unit itself to turn your weaker topics into strengths.

Your **exam technique** will improve if you attempt the concept questions, and then check the answers. Next, practice with past exam papers from your exam board. You should be able to download them from the exam board with the answers.

HOW TO SUCCEED IN EXAMS

When you understand the concepts and have memorised any necessary ideas, you can work on improving your **exam technique**.

To gain marks in exams you must, of course, understand and know the topics. You must also have an effective exam technique; remember to read the questions carefully, make sure you understand what the examiners are asking, answer the question (rather than just writing anything you know), and communicate a clear answer using technical words correctly.

Students commonly lose marks by not reading the questions. It may appear obvious, but in the stress of the exam many students do not always read every word and so do not answer a question appropriately. So always read the question at least twice. Many students write everything they know about the subject mentioned in the question, and so produce long-winded answers. This may waste so much time that they do not finish the exam paper. More importantly, examiners state that long rambling answers tend not to gain full marks because they do not focus on the particular point raised in the question.

The examiners use particular 'command words' which indicate how you should respond to questions; marks are easily lost by ignoring these. For example, many students 'explain' when the question says 'describe'. Here is a list of command words found in A-level chemistry exam papers.

Define or *What is meant by...* – Give a definition in words and in equations if possible.

Describe – This asks you to state what is observed in an experiment, or state the basic points in a practical method. Giving a chemical explanation is not necessary.

Describe what will be observed – 'Observed' means seen, or sensed, so describe colours, states, and smells. Chemical names may not gain you marks.

Explain – say how and why something happens. Be careful to use the correct technical words. If many marks are offered then explain in depth.

Calculate – Obviously, work out a numerical answer. Remember to give the correct sign (for example, exothermic enthalpies are negative), and give your answer to the correct number of significant figures.

Using the data given – You must refer to the data given! Show the examiner you have done so by marking graphs, using figures in calculations, or using words from the question. Be wary of basing your answer on recall of knowledge.

Give the formula – You must give the formula, not the name. Easily overlooked.

Name – Give the name. The examiner is checking that you can name compounds. A formula will not do.

Identify – Give the chemical name or formula.

Suggest – Anything reasonable will do as an answer. There are many possible responses. This confuses some students because it is so unusual in chemistry exams; the examiner usually wants a particular answer.

Comment on – The examiner wants you to point out an idea, usually in the specification (syllabus), suggested by the data.

Here are common technical words that students misuse. Take care not to confuse them. Examiners will not give marks if you talk about atoms in sodium chloride – because it contains ions, of course.

Words for particles are *atom*, *ion*, and *molecule*.

Chemical substances are *elements*, *compounds*, or *mixtures*.

Bonding must be *covalent, ionic,* or *metallic*...

...Whereas structures must be *simple covalent molecules* or *giant lattices* (which could be *covalent, ionic,* or *metallic*).

THE A LEVEL SYSTEM

The system is designed so that a typical student will study four or five ASs in their first year of study, and then select three from these to continue in their second year of study, called A2. This will then give them three full A levels. The system is designed to be flexible, so that it is possible to do any number of ASs and A2s, but an A2 can only be done if the relevant AS has been completed. In reality, most students' choices will be limited by whatever system and options their school or college can offer.

Two of the three main examination boards are offering two specifications (syllabuses) in chemistry, and one main board is offering one specification. Each AS component contains three modules, as does each A2, but the content of each module varies from specification to specification. Whatever specification a student studies, most of the content covered is the same as in any other specification, but the topics are mixed differently in the different modules.

OCR EXAM BOARD SPECIFICATIONS AND MODULES

OCR Exam Unit weighting	F324, 90 marks A2 30%, A level 15%		F325, 90 marks A2 50%, A level 25%		F326, 60 marks A2 20%, A level 10%
	Book Unit	OCR code	Book Unit	OCR code	
Links	1	4.1.1	7	5.1.1	book unit 15 covers
	2	4.1.2	8	5.1.2	the practical
	3	4.13, 4.14	9	5.1.3	coursework component
	4	4.2.1, 4.2.2	10	5.2.1	
	5	4.2.3	11	5.2.2	
	6	4.3.1, 4.3.2	12	5.23	
			13	5.3.1	
			14	5.3.1	

The specifications may be downloaded from the relevant exam board. They are useful when revising. The main OCR website is:
http://www.ocr.org.uk/index.html

THE NEW A* GRADE

To gain a grade **A*** you must:
Gain a **grade A, 80%, at A level** which is at least 480/600 UMS.
Gain at least **90% in the A2** units overall which is at least 270/300 UMS.

In this book, A* style questions have been included in the 'Concept Test' section in each unit.

Unit 1 ARENES

Aromatic chemistry is the study of compounds based on the **benzene ring** C_6H_6.

- Compounds based on the **benzene ring** are called **arenes**. The benzene molecule is a ring of six carbon atoms; each carbon atom is joined to two others and a single hydrogen atom. You will meet the structure drawn in a number of ways (see Fig. 1.1).

Fig. 1.1

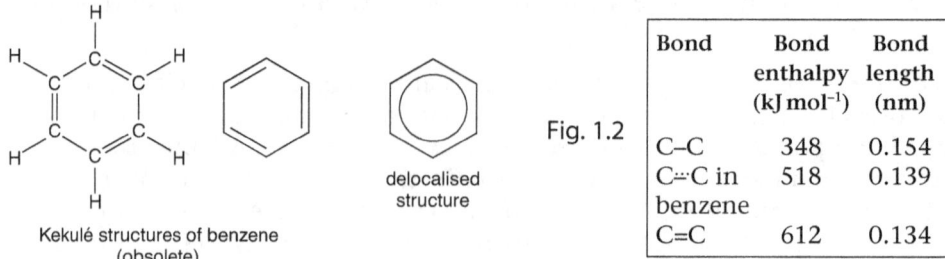

Fig. 1.2

Bond	Bond enthalpy (kJ mol⁻¹)	Bond length (nm)
C–C	348	0.154
C⋯C in benzene	518	0.139
C=C	612	0.134

AS organic chemistry is necessary for this topic.

- Benzene is unsaturated, but it does not react with bromine (by addition) like an alkene. It is very unreactive. The six carbon–carbon bonds are of equal length, not alternating single and double bonds (see Fig. 1.2).

If you compare the **hydrogenation enthalpy** for **benzene** with the hydrogenation enthalpy for **three cyclohexene molecules** (3 × C=C), then you will see the hydrogenation of benzene releases less heat (see Fig. 1.3). The **difference in the enthalpies** of hydrogenation represents the **extra stability** of the delocalised electron structure.

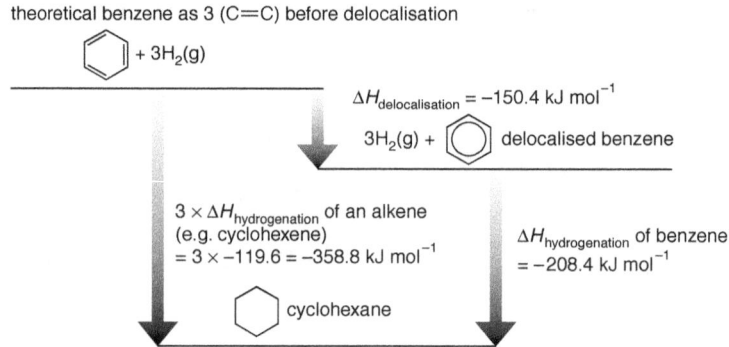

- The **p atomic orbitals** on each carbon atom overlap with each other to form a set of **delocalised π bonds** (see Fig. 1.4). The six delocalised electrons in these orbitals have lower energy than the six electrons in 3 C=C bonds.

In contrast the electron density of the π bonds of the C=C in alkenes are localised so alkenes easily reacts with bromine, whereas the delocalised electron density of the π bonds in benzene are delocalised so benzene is resistant to bromination.

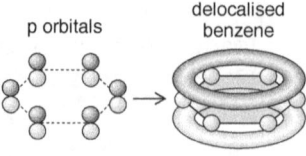

Fig. 19.4

- The stability of the electrons delocalised around the benzene ring means that it does not undergo electrophilic addition. Rather, it undergoes **electrophilic substitution** of the hydrogen atoms by strong electrophiles.

Typical electrophilic substitution reactions include:
Nitration to form nitrobenzene: $C_6H_6 + HNO_3$ (in H_2SO_4) $\to C_6H_5NO_2 + H_2O$

Concentrated HNO_3 (in H_2SO_4) is the 'nitrating mixture' that provides the nitronium ion electrophile NO_2^+ (see Fig. 1.5).

| AlCl₃ is called a **halogen carrier** in halogenation reactions. |

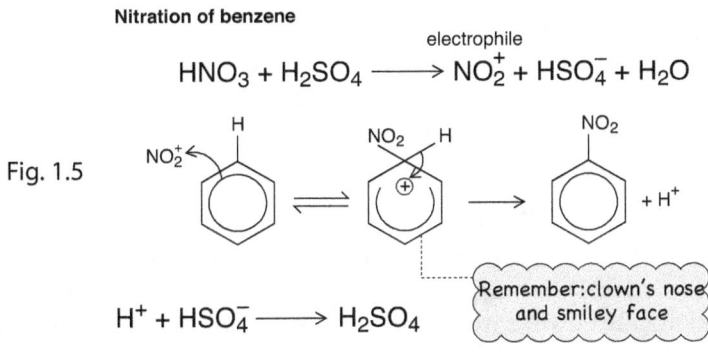

Nitration of benzene

$$HNO_3 + H_2SO_4 \longrightarrow NO_2^+ + HSO_4^- + H_2O$$

Fig. 1.5

$$H^+ + HSO_4^- \longrightarrow H_2SO_4$$

Remember: clown's nose and smiley face

| Other halogen carriers include $FeCl_3$, $AlBr_3$, and $FeBr_3$. |

| Nitrated aryl compounds can be used as explosives e.g. 'trinitrotoluene' TNT $CH_3C_6H_2(NO_2)_3$. |

| **Nitrobenzene** $C_6H_5NO_2$ may be reduced to **phenylamine** $C_6H_5NH_2$ which is then converted into azo dyestuffs. |

Alkylation to form e.g. ethylbenzene (the 'Friedel–Crafts' reaction):

$C_6H_6 + CH_3CH_2Cl \xrightarrow{AlCl_3} C_6H_5CH_2CH_3 + HCl$ (see Fig. 1.6).

Alkyation of benzene

$$CH_3Cl + AlCl_3 \longrightarrow CH_3^+ + AlCl_4^-$$

Fig. 1.6

$$H^+ + AlCl_4^- \longrightarrow AlCl_3 + HCl \text{ regenerated}$$

Clown's nose and smiley face

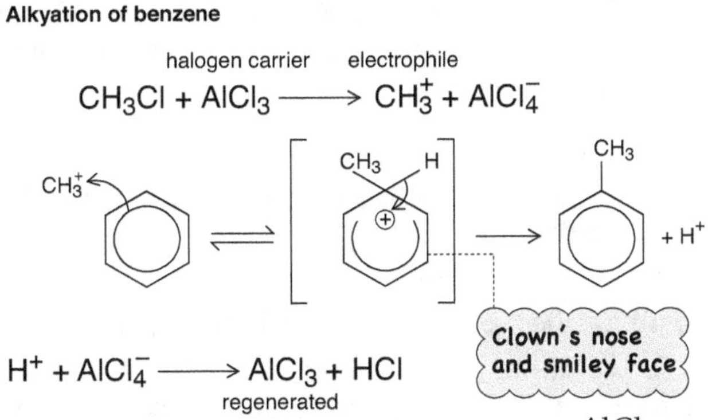

Chlorination to form chlorobenzene: $C_6H_6 + Cl_2 \xrightarrow{AlCl_3} C_6H_5Cl + HCl$

Bromination to form bromobenzene: $C_6H_6 + Br_2 \xrightarrow{AlBr_3} C_6H_5Br + HBr$

Acylation to form e.g. phenylethanone:

$C_6H_6 + CH_3COCl \xrightarrow{AlCl_3} C_6H_5COCH_3 + HCl$

- Note that, except for nitration, **AlCl₃** is used as a **catalyst**. It must be used in **anhydrous** conditions: water would hydrolyse it. $AlCl_3$ forms dative (co-ordinate) bonds with the other reagent to produce the positively **charged electrophile** e.g.

  Chlorination: $AlCl_3 + Cl_2 \to AlCl_4^- + \mathbf{Cl^+}$
  Alkylation: $AlCl_3 + CH_3CH_2Cl \to AlCl_4^- + \mathbf{CH_3CH_2^+}$
  Acylation: $AlCl_3 + CH_3COCl \to AlCl_4^- + \mathbf{CH_3CO^+}$

  You can **brominate** benzene with a mixture of **Fe** and **Br₂**. Some of the bromine reacts with the iron to make iron bromide which then acts as the halogen carrier for bromination.

- The **-CH₃** group on methylbenzene, like all alkyl groups, will be converted into **-CH₂Cl** by **homolytic free radical substitution** using chlorine and ultraviolet light (see Fig. 1.7). Contrast this reaction with chlorination of the benzene ring itself, using Cl_2 and $AlCl_3$.

Fig. 19.7

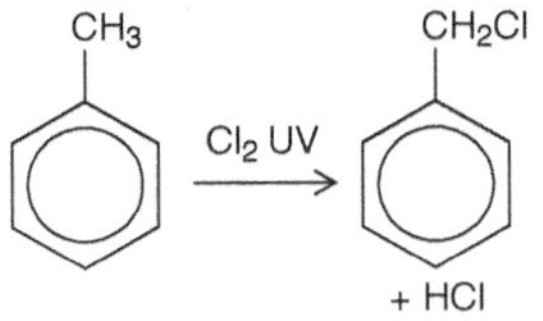

CH_3-C₆H₅ $\xrightarrow{Cl_2 \text{ UV}}$ CH_2Cl-C₆H₅ + HCl

Unit 1

ARENES

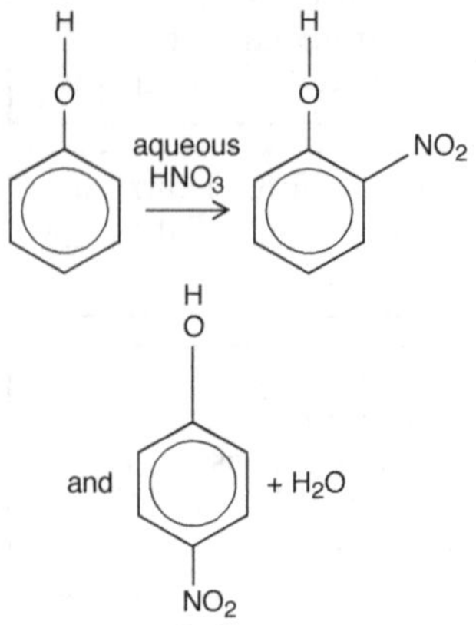

- **Phenol** is used in the production of plastics, antiseptics, disinfectants and resins for paints.

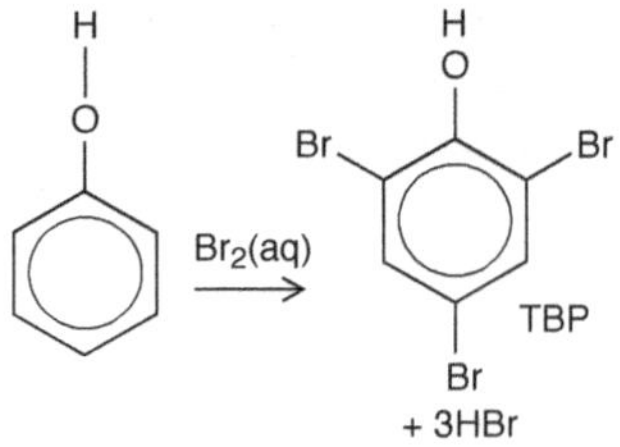

  Phenol C_6H_5OH is **acidic** due to electron-pair donation to the delocalised ring from the oxygen p-orbital, which allows the H^+ to leave.

  Phenol is a very **weak acid**. It does turn blue litmus to red, but does *not* liberate CO_2 from carbonates. It will react with NaOH(aq):

  $C_6H_5OH(aq) + NaOH(aq) \rightarrow C_6H_5O^-Na^+(aq) + H_2O(l)$

  Like all alcohols, phenol reacts with sodium:

  $C_6H_5OH(aq) + Na(s) \rightarrow C_6H_5O^-Na^+(aq) + H_2(g)$

  The phenol group activates the ring to electrophilic substitution, making phenol much **more reactive** than benzene. Again this is due to electron-pair donation to the delocalised ring from the oxygen p-orbital. For example, phenol is nitrated by aqueous nitric acid (see Fig. 1.8) and is brominated by aqueous bromine to make 2,4,6-tribromophenol (TBP), a white suspension (see Fig. 1.9). In a similar way, the antiseptic TCP, 2,4,6-trichlorophenol, is made from phenol and chlorine.

  In common with alcohols, phenol will form esters (see unit 3).

  The **test** for a **phenolic group** is to add aqueous iron III chloride. Any phenolic group will make a particularly evil-looking violet solution (an iron III phenolic complex).

- **Nitrobenzene** may be converted into **phenylamine** by reduction using tin with concentrated hydrochloric acid.

  $C_6H_5NO_2 + 6[H] \rightarrow C_6H_5NH_2 + 2H_2O$

  In common with all amines, phenylamine is basic (see unit 3).

  Phenylamine is converted to a **diazonium ion** using nitrous acid HNO_2 (or HONO). This reagent is unstable so must (i) be kept cold and (ii) be made when needed by mixing a source of nitrite ions (NO_2^-) with H^+ ions, e.g.

  $NaNO_2(aq) + HCl(aq) \rightarrow HNO_2(aq) + NaCl(aq)$

- You must write down the production of **nitrous acid** and the **diazotisation reaction** as shown in Fig. 1.10.

- Ice cold aqueous **phenol** with alkali converts the diazonium ion into an **azo dye** 4-hydroxyphenylazobenzene. Two molecules join together in this **coupling reaction** (see Fig. 1.10).

Fig. 1.8

Fig. 1.9

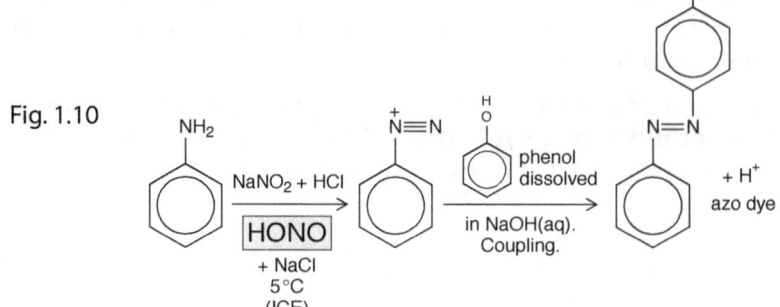

Fig. 1.10

> **Extensively delocalised** structures are often coloured. The molecular structure responsible for the colour is called the **chromophore**. (See also 2,4-DNP.)

TESTS

RECALL TEST

1. Explain why all the carbon–carbon bonds in benzene have the same length, and a length intermediate between the C-C and C=C bond lengths.

2. Explain how the delocalisation energy may be determined.

3. Draw the mechanism for nitration of benzene.

4. Draw the mechanism for forming methylbenzene from benzene.

5. Write balanced equations for:

    a benzene + concentrated nitric acid (with H_2SO_4(l))

    b benzene + chlorine (with $AlCl_3$)

    c benzene + bromine (with $FeBr_3$)

    d benzene + ethanoyl chloride (with $AlCl_3$)

6. Explain why aluminium chloride must be anhydrous for electrophilic substitution to occur.

Unit 1 — TESTS

7 Suggest how can styrene (phenylethene) be made from benzene.

8 Give two uses for nitrated aromatic compounds.

9 State what is made when sodium is added to phenol.

10 Give the reagents and conditions required to convert:
   a benzene into nitrobenzene

   b nitrobenzene into phenylamine

   c phenylamine into a diazonium salt

   d a diazonium salt into an azo dye

   e phenol into sodium phenoxide

   f phenol into 2,4,6-trichlorophenol

CONCEPT TEST

1 Benzene may be converted into methylbenzene.
   a Draw the mechanism for this reaction.

   b Name this mechanism.

   c Why is cyclohexene more reactive than benzene?

   d Why is the bromination of phenol much easier than the bromination of benzene?

2 Methylbenzene, A, may be converted by a sequence of reactions to produce a very useful compound, E.
First methylbenzene, A, is nitrated using concentrated nitric and sulphuric acids to produce substance B.
Then B is converted into compound C, with the molecular formula C_7H_9N.
C reacts with aqueous sodium nitrite and sulfuric acid, in ice, to form a solution containing the ion D, with the empirical formula $C_7H_7N_2^+$.
Solution D, when mixed with an alkaline phenol solution, in ice, forms a brightly coloured solid, E.
If the solution D were allowed to warm up then a phenol would form.

   a   Give the structure of B.

   b   Give the reagents required to convert B into C.

   c   Give the structure of the ion D.

   d   What type of substance is E?

   e   Draw the structure of E.

   f   Give the test for phenol.

3 Give the reagents and conditions necessary to carry out the following changes:

   a C_6H_6 to $C_6H_5CH_3$

   b $C_6H_5CH_3$ to $C_6H_5CH_2Cl$

   c $C_6H_5CH_2Cl$ to $C_6H_5CH_2OH$

   d $C_6H_5CH_3$ to $NO_2C_6H_5CH_3$

   e C_6H_5OH to $C_6H_5O^-Na^+$

Unit 2 — CARBONYL COMPOUNDS

AS organic chemistry is necessary for this topic.

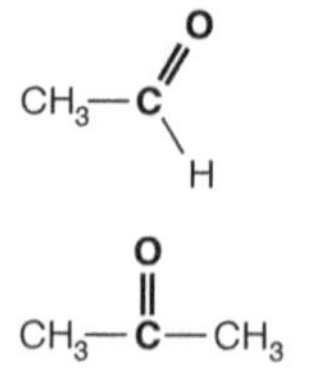

ethanal: CH_3-CHO
propanone: $CH_3-CO-CH_3$

Fig. 2.1

Aldehydes are reduced to primary alcohols.

Ketones are reduced to secondary alcohols.

- Aldehydes and ketones are **carbonyl compounds** because they contain the >C=O carbonyl group (see Fig. 2.1). The C=O bond is **polarised** and gives each molecule a permanent dipole (but no H-bonding). **Dipole–dipole interactions** give aldehydes and ketones **higher b.p.s** than alkanes of corresponding relative molecular mass.

  Ethanal CH_3CHO (b.p. 21 °C) and **propanone** CH_3COCH_3 (b.p. 56 °C) **burn** in oxygen with a blue flame:

  $2CH_3CHO + 5O_2 \rightarrow 4CO_2(g) + 4H_2O(l)$

  $CH_3COCH_3 + 4O_2 \rightarrow 3CO_2(g) + 3H_2O(l)$

  Aldehydes smell like **apples** and are toxic.
  Ethanal (made from ethene) is the **starting material** for many organic compounds. As an oxidation product of ethanol, it is responsible for hangovers.
  Ketones have a **sweet smell** and are also toxic.
  Propanone (acetone), an important **solvent**, was once used as nail varnish remover.

- The **reduction** of aldehydes and ketones gives **alcohols**. $NaBH_4$(aq) or $LiAlH_4$ (dry ether) reduce **aldehydes** to **primary** alcohols and **ketones** to **secondary** alcohols. Use [H] to indicate a reducing agent.

  Example: Ethanal reduced to ethanol. $CH_3CHO + 2[H] \rightarrow CH_3CH_2OH$

  Example: Propanone reduction: $CH_3COCH_3 + 2[H] \rightarrow CH_3CH(OH)CH_3$

  Reduction occurs because the **H⁻ ion** acts as a **nucleophile** (see Fig. 2.2) by **nucleophilic addition**. Note that H⁺ ions (protons) are also needed, supplied either by water or by the later addition of aqueous acid. The H⁺ ion protonates the organic intermediate.

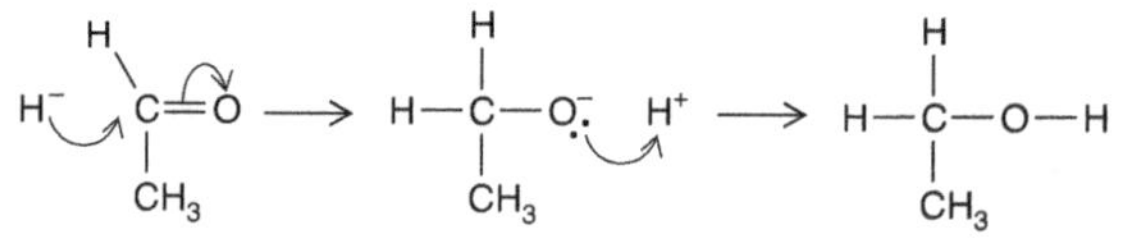

  Fig. 2.2

- Most common **oxidising agents** (e.g. acidified aqueous dichromate) will readily oxidise **aldehydes** to **carboxylic acids**. Ketones resist oxidation.
  Example: $CH_3CHO + [O] \rightarrow CH_3COOH$

- Use Tollens' reagent to **distinguish** between **aldehydes** and **ketones**. Only aldehydes are oxidised by these reagents and give a positive result.

  Tollen's reagent ('ammoniacal silver nitrate') contains **Ag⁺** ions which are reduced during oxidation of the aldehyde to a distinctive **silver mirror** on the inside of the test tube (or else a black precipitate of silver). In reactions showing the action of Tollen's [O] is acceptable, so the reduction of propanal by Tollen's may be written:

  $CH_3CH_2CHO + [O] \rightarrow CH_3CH_2COOH$

  The reduction of the silver ions may be shown as:

  $Ag^+(NH_3(aq)) + e^- \rightarrow Ag(s)$

  or as a complex (see Transition Metals, unit 15):

  $[Ag(NH_3)_2]^+(aq) + e^- \rightarrow Ag(s) + 2NH_3(aq)$

To make **Tollen's reagent**: (i) Add aqueous **NaOH** to aqueous **silver nitrate** to make a pale brown precipitate of silver hydroxide; (ii) drip aqueous **ammonia** into the solution until the precipitate disappears. To **test** for an aldehyde, add the organic substance to this solution and warm gently. If an aldehyde is present a silver mirror (or black precipitate) will form.

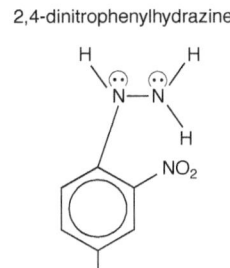

Fig. 2.4

- Aldehydes and ketones react by **nucleophilic addition**, i.e. heterolytic nucleophilic addition occurs when HX reacts with any carbonyl compound.
  $CH_3COCH_3 + HX \rightarrow CH_3COHXCH_3$

For another example of **nucleophilic addition** see Fig. 2.3 for the mechanism of **HCN addition**. You will not have to recall this reaction, but you may have to suggest a similar mechanism in an analytical A2 question.

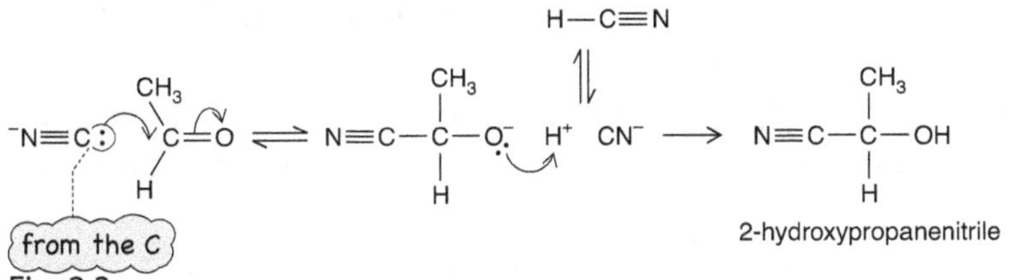

Fig. 2.3

Addition of **HCN** to the C=O group produces **two functional groups**: a hydroxy group **-OH** and a nitrile **-CN**.

- To **test** for aldehydes and ketones, you add **2,4-dinitrophenylhydrazine** (2,4-DNP) (see Fig. 2.4) to make **dinitrophenylhydrazones**, which are brightly coloured **yellow-orange solids**. 2,4-dinitrophenylhydrazine reacts by a **condensation** reaction.

  Example: Ethanal to make **ethanal 2,4-dinitrophenylhydrazone**.

  You must know the full name of 2,4-dinitrophenylhydrazine but not know its structure (see Fig. 2.5). It helps to remember that **hydrazine** is NH_2NH_2 and that **phenyl** indicates a benzene ring; then add two **nitro groups** (-NO_2) at positions **2** and **4** on the ring. You do not have be able to draw the structures of dinitrophenylhydrazones.

- Each dinitrophenylhydrazone, purified by recrystallization, has a **distinctive melting point** that indicates which carbonyl compound was originally present (see Fig. 2.5 for melting point apparatus).

- Use **recrystallization** to purify a compound which is more soluble in hot solvent than cold. The stages are:
  (i) **Dissolve** the impure substance (which is effectively a mixture) in the **minimum** amount of **hot solvent** (inorganic compounds in water; organic compounds in water, or ethanol, or a liquid hydrocarbon);
  (ii) **Filter** the hot mixture under reduced pressure, to remove insoluble impurities (see above);
  (iii) **Cool** the filtrate slowly to form crystals (fast cooling traps impurities within the crystals);
  (iv) **Filter** the cold mixture, to remove soluble impurities. The **residue** is the pure substance.

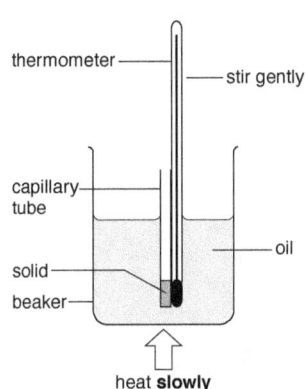

Fig. 2.5 Melting point apparatus

Unit 2

EXAMPLES

- You must recall the details studied in unit 2, so below, for your convenience, is the AS organic content that includes carbonyl compounds.

- A molecule is **oxidised** when it **gains O** or **loses H**.
  Example: Atmospheric oxygen oxidises the alcohol CH_3CH_2OH (ethanol) to vinegar CH_3COOH (ethanoic acid). The alcohol molecule has lost two H atoms and gained one O.

- A molecule is **reduced** when it **loses O** or **gains H**.
  Example: Ethanal CH_3CHO gains two H atoms when it is reduced to ethanol CH_3CH_2OH. A powerful **reducing agent** is needed.

- Alcohols may be **oxidised** by combustion. Ethanol burns with a clean blue flame. $CH_3CH_2OH + 3O_2 \rightarrow 2CO_2 + 3H_2O$

- When exposed to the air, **primary alcohols** RCH_2OH will **oxidise** very slowly to **aldehydes** RCHO and then to carboxylic acids RCOOH (e.g. beer and wine slowly change to vinegar).

- The strong oxidising agents acidified potassium dichromate(VI) $K_2Cr_2O_7$ and acidified potassium permanganate manganate(VII) $KMnO_4$ are used in the lab.

- To stop the oxidation at the **aldehyde**, the oxidising agent $K_2Cr_2O_7$ with H_2SO_4 is dripped into hot **primary alcohol**, and the aldehyde is heated and **distilled** off as it forms. (See Fig. 2.6)
  $CH_3CH_2OH + [O] \rightarrow CH_3CHO + H_2O$

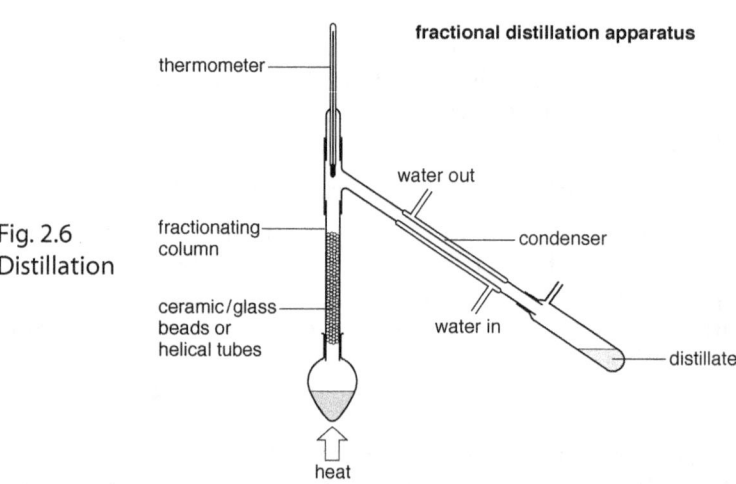

Fig. 2.6 Distillation

The aldehyde has a polar C=O so has a lower b.p. (40°C) than the hydrogen-bonded alcohol (76°C). Use $NaBH_4(aq)$ to reduce back to the alcohol.
$CH_3CHO + 2[H] \rightarrow CH_3CH_2OH$

- To oxidise a primary alcohol to the carboxylic acid, heat under reflux with the oxidising agent (see Fig. 2.7) and then distil off the product.
  $CH_3CH_2OH + 2[O] \rightarrow CH_3COOH + H_2O$
  (Use $LiAlH_4$ to reduce back to the alcohol.)

- **Secondary alcohols** oxidise to form a **ketone** e.g. propan-2-ol plus oxidising agent form propanone and water.
  $CH_3CH(OH)CH_3 + 2[O] \rightarrow CH_3COCH_3 + H_2O$

- **Tertiary alcohols** cannot easily be oxidised except by combustion, $(CH_3)_3COH$ resists oxidation.

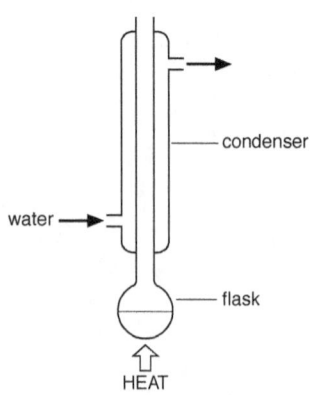

Fig. 2.7 Heat under reflux

TESTS

RECALL TEST

1 What intermolecular forces do aldehydes and ketones have?

2 Both aldehydes and ketones often react by the mechanism _____ _____. Both may be reduced to form an _____. Only _____ _____ may be oxidised to a _____.

3 Draw the mechanism for the reaction between propanone and $NaBH_4$(aq).

4 Which has the lower boiling point, propanal or propan-1-ol? Explain your answer.

5 Name the product of the reaction between ethanal and air.

6 a State the test which is positive for both ketones and aldehydes:

Reagent

What happens?

b How could a carbonyl compound be identified from the solid derivative?

7 a How could propanal be prepared from an alcohol? State the alcohol required.

b How could propanone be prepared from an alcohol? State the alcohol required.

Unit 2 TESTS

8 State the reaction type and the product formed when NaBH$_4$(aq) is reacted with:

    a propanal

    b propanone

9 Write a balanced equation for these reactions.

    a NaBH$_4$(aq) with propanal

    b LiAlH$_4$(dry ether) with propanone

    c Ammoniacal silver nitrate with ethanal

    d ethanal with an oxidising agent

10 Name the test that is positive with aldehydes but not ketones.

11 Propanone may be converted into propan-2-ol. State the required reagents and conditions.

12 Give the reagents and conditions for these conversions:

    a ethanol to ethanoic acid

    b ethanol to ethanal

    c propan-2-ol to propanone

CONCEPT TEST

1 Unknown compound X, C$_3$H$_8$O, is heat with acidified K$_2$Cr$_2$O$_7$ to form compound Y which reacts with 2,4-dinitrophenylhydrazine to produce a yellow solid. Y does not react with silver nitrate that is dissolved in dilute ammonia. Compound X was made from compound Y,

    a What type of compound is Y? Explain your reasoning.

    b Deduce the formula of compound Y. Explain your reasoning.

c Deduce the formula of compound X. Explain your reasoning.

d How could compound Y be converted back into compound X?

2 a Give a test that will distinguish between an aldehyde and a ketone.

b How may the 2,4-dinitrophenylhydrazine derivative be purified?

c Explain how the pure derivative may be used to confirm the identity of a carbonyl compound.

3 To answer this question you must have studied spectroscopy.

a Propanal and propanone have the same relative molecule mass. How could this be determined? Describe the results produced.

b The infra red spectra of propanal and propanone have similar characteristics. What feature do all carbonyl compound infra red spectra show?

c The proton NMR of propanal and propanone very different.
i Which chemical shift do both carbonyl compounds produce?

ii Which molecule will only produce one chemical shift? Explain why only on peak is made and why it is not split into sub-peaks.

iii State and explain the chemical shift, if any, due to the CO group.

iii Explain why the -CH_2- group in propanal splits into sub-peaks. How many peaks will be seen within the peak due to the -CH_2- group?

Unit 3 — CARBOXYLIC ACIDS, ESTERS AND AMINES

AS organic chemistry is necessary for this topic.

- **Carboxylic acids** contain the functional group **-COOH** (see Fig. 3.1). The C=O and the O-H bonds are **polar**, giving extensive **H-bonding** as well as **dipole-dipole** attraction between the molecules so carboxylic acids have high melting and boiling points.

Fig. 3.1 Ethanoic acid forms hydrogen bonds. It also forms dimerises.

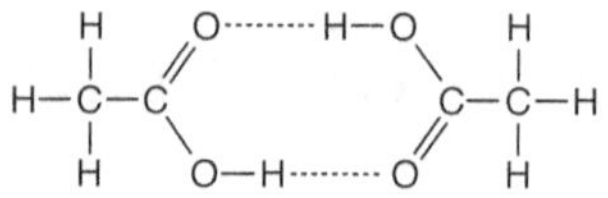

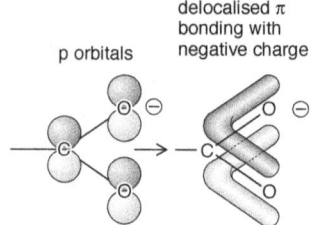

Fig. 3.2

The **carboxylic acid group** is written as **COOH** (not CO_2H) to show that the oxygen atoms are different.

The **carboxylate ion** is written CO_2^- (not COO^-) to show that the two oxygen atoms are equivalent.

When pure liquids, or in non-polar solvents, carboxylic acids **dimerise**. Two molecules H-bond to form a single **non-polar dimer**.

Carboxylic acids form stable $-CO_2^-$ ions because of **delocalisation** of the negative charge around the three atomic centres. The C and O atoms all have similar sizes which allows the p orbitals to overlap easily (see Fig. 3.2).

- Carboxylic acids **dissociate incompletely** in aqueous solution and so are **weakly acidic**. The sour taste of vinegar that improves the flavour of fish and chips is due to the formation of $H^+(aq)$ ions.

$CH_3COOH(aq) \rightleftharpoons CH_3CO_2^-(aq) + H^+(aq)$

Remember that the alternative representation is:

$CH_3COOH(aq) + H_2O(l) \rightleftharpoons CH_3CO_2^-(aq) + H_3O^+(aq)$

The **strength** of a carboxylic acid depends on how easily H^+ can leave, which depends on the **electron density** of the O-H group. If there is an electron-donating group (e.g. $-CH_3$), causing a positive inductive effect (+I), then the O-H bond is electron rich; it is more difficult for H^+ to leave and the acid is **weaker**. If the group is **electron withdrawing** (e.g. -Cl or a benzene ring), giving a negative inductive effect (–I), then the O-H bond is electron deficient; it is easier for H^+ to leave and the acid is **stronger**. The inductive effect explains the increase in acidity: $CH_3COOH < C_6H_5COOH < HCOOH$.

- **Carboxylic acids** react as typical acids. They turn litmus from **blue to red** and **neutralise bases**. Example:

$CH_3COOH(aq) + NaOH(aq) \rightarrow CH_3CO_2^-Na^+(aq) + H_2O(l)$

They liberate **carbon dioxide** (with effervescence) from carbonates and hydrogencarbonates. Example:

$CH_3COOH(aq) + NaHCO_3(s) \rightarrow CH_3CO_2^-Na^+(aq) + H_2O(l) + CO_2(g)$

Like most acids they also oxidise reactive metals. Example:

$CH_3COOH(aq) + Na(s) \rightarrow CH_3CO_2^-Na^+(aq) + H_2(g)$

- Carboxylic acids can be **reduced** to **alcohols**. The powerful reducing agent $LiAlH_4$ (in dry ether) **must** be used. Example:

$CH_3COOH(aq) + 4[H] \rightarrow CH_3CH_2OH(l) + H_2O(aq)$

- Carboxylic acids react with **alcohols** to make **esters** and water (see Fig. 3.3). Example:

$CH_3COOH(aq) + CH_3CH_2OH(aq) \rightleftharpoons CH_3COOCH_2CH_3(aq) + H_2O(l)$

Fig. 3.3

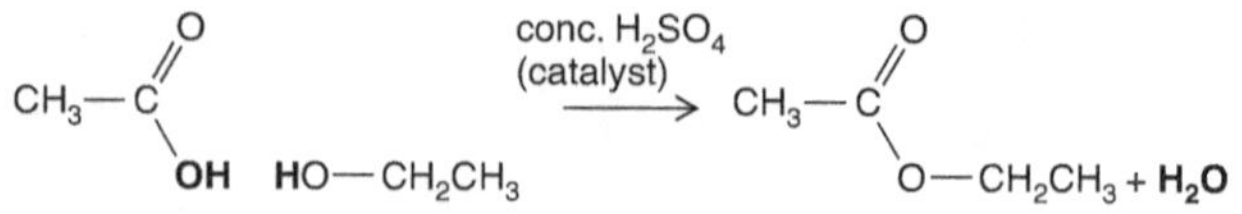

- Esters are also produced by reacting dry **acid anhydrides** and **alcohols** to make **esters** and a carboxylic acid. Example:

$(CH_3CO)_2O(l) + CH_3CH_2OH(aq) \rightleftharpoons CH_3COOCH_2CH_3(aq) + CH_3COOH(l)$

- Many esters have **pleasant smells**. Fruity ethyl ethanoate is used as a food flavouring (e.g. in sweets) and as an ink solvent in some 'spirit' felt marker pens. Some esters are used in perfumes, as solvents, or as flavourings.

    Esters are **hydrolysed** by heating under reflux with **dilute aqueous acid** (e.g. H_2SO_4) to form a **carboxylic acid** and an **alcohol**. (The $H^+(aq)$ ions catalyse the reaction.) **Example:**

    $CH_3COOCH_2CH_3(l) + H_2O(l) \rightleftharpoons CH_3COOH(l) + CH_3CH_2OH(l)$

    Esters are also **hydrolysed** by heating under reflux with dilute alkali (aqueous base) to form the **salt** of the carboxylic acid and an **alcohol**. (The $OH^-(aq)$ ions catalyse the reaction.) **Example:**

    $CH_3COOCH_2CH_3(l) + NaOH(aq) \rightarrow CH_3CO_2^-Na^+(aq) + CH_3CH_2OH(aq)$

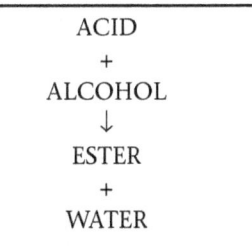

- **Fats** and **oils** are **triglyceride esters or 'fatty esters'** (esters of the trihydric alcohol, glycerol which is propane-1,2,3-triol, $CH_2OHCHOHCH_2OH$). Their alkaline hydrolysis (which is called **saponification**) is used to manufacture **soap**.

    Animal fats and palm oil are saturated fatty esters. These molecules fit together and so form solids that cause heart disease.

    Most vegetable oils are unsaturated and so form E-Z isomers called 'cis' and 'trans' fatty esters. The 'cis' form is common in vegetable oils. The cis form is converted to the 'trans' form by heating.

    The systematic names for fats and oils use a number notation. Olive oil contains 'octadec-9-enoic acid, 18,1', where 'octadecenoic acid' indicates a carboxylic acid with 18 carbons with the 'en' indicating that there is a C=C. The '9' indicates the C=C is nine carbons along the chain with carbon number one being in the -COOH. The '18,1' is short hand for a 18 carbon carboxylic acid, with 1 C=C which is at the 9th position. **Example:** octadecanoic acid,18,0; also octadec-9,12-enoic acid, 18,2(9,12)

    Saturated fats and **'trans' unsaturated** vegetable oils are linked to 'bad' cholesterol and the resultant increased risk of coronary heart disease and strokes.

    Fatty esters are increasingly used as **biodiesel** which is a renewable resource. However there is concern that some biodiesel production depends on the destruction of rainforest and actually increases carbon dioxide emissions due to the burning of soils and the use of synthetic fertilisers.

- **Acid anhydrides** are made from carboxylic acids using a dehydrating agent:

    $2CH_3COOH(l) \rightarrow (CH_3CO)_2O(l) + H_2O(l)$

- **Acid anhydrides** (see Fig 3.4) react like **very reactive carboxylic acids**, so only a little warming a required.

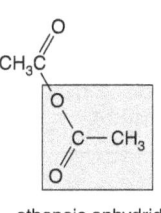

ethanoic anhydride

Fig. 3.4

- Acid anhydrides react with the **nucleophile** such as water (forming **carboxylic acids**), and alcohols (forming **esters**). Examples:

    With water: $(CH_3CO)_2O(l) + H_2O(l) \rightarrow 2CH_3COOH(aq)$

    With alcohol:

    $(CH_3CO)_2O(l) + CH_3CH_2OH(l) \rightarrow CH_3COOCH_2CH_3(aq) + CH_3COOH$

- **Aspirin** is an ester made from 2-hydroxybenzoic acid and ethanoic anhydride. The by-product is the non-toxic ethanoic acid.

    $(CH_3CO)_2O(l) + HOC_6H_4COOH(s) \rightarrow CH_3COOC_6H_4COOH(s) + CH_3COOH(l)$

With ethanol,

acid anhydrides are fairly reactive (a little heat helps),

carboxylic acids are the least reactive (heat under reflux).

Unit 3 — CARBOXYLIC ACIDS, ESTERS AND AMINES

AS organic chemistry is necessary for this topic.

- Amines contain the **-NH₂** functional group. They can act as proton acceptors (bases) and as **nucleophiles**. The polar N-H group enable amines to **H-bond**, so they have high m.p.s and b.p.s.

$CH_3CH_2NH_2$
This is aminoethane, also called ethylamine.

- The -NH₂ lone pair can attract a proton (H⁺ ion) and bond to it:
  $R-NH_2 + H^+ \rightarrow R-NH_3^+$

  Amines are basic so they react with **acids** to make **salts**. **Example:** ethylamine forms the salt ethyl ammonium chloride.
  $CH_3CH_2NH_2 + HCl \rightarrow CH_3CH_2NH_3^+Cl^-$
  Example: ethylamine also forms the salt ethyl ammonium sulfate.
  $CH_3CH_2NH_2 + H_2SO_4 \rightarrow (CH_3CH_2NH_3)SO_4$

- In common with all **amines**, phenylamine is **basic**. It has limited solubility in water but **dissolves** readily in acid to form the aqueous **phenylammonium ion**.
  $C_6H_5NH_2(l) + H^+(aq) \rightarrow C_6H_5NH_3^+(aq)$
  which will react with alkali to **regenerate** the free base.
  $C_6H_5NH_3^+ + OH^-(aq) \rightarrow C_6H_5NH_2(l) + H_2O(l)$

Primary aliphatic amines are best prepared by reacting alcoholic ammonia to a haloalkane. **Aromatic amines** are prepared by **reducing aromatic nitro** compounds using tin and concentrated HCl.

- Ammonia (dissolved in ethanol) will **react repeatedly** with a **haloalkane** by nucleophilic substitution to make various amines. Example:
  (i) $NH_3 + CH_3Br \rightarrow HBr + CH_3NH_2$ (a **primary** amine)
  (ii) $CH_3NH_2 + CH_3Br \rightarrow HBr + (CH_3)_2NH$ (a **secondary** amine)
  (iii) $(CH_3)_2NH + CH_3Br \rightarrow HBr + (CH_3)_3N$ (a **tertiary** amine)
  (iv) $(CH_3)_3N + CH_3Br \rightleftharpoons Br^- + (CH_3)_4N^+$ (a **quaternary** ammonium ion)
  Long-chain quaternary ammonium ions are used as **detergents** (cationic surfactants) in bubble bath liquids and washing powders. They are also **mildly antiseptic** and are used in throat lozenges (look on ingredients labels).

- Aromatic amines may be made into **azo dyes**. See Unit 1 for details.

- Only students striving for an A* should learn this:

  The **attractive power** of the lone pair determines the -NH₂ base strength. If the -NH₂ group is joined to an **electron-donating group** (the +I inductive effect of e.g. -CH₃ or -CH₂CH₃, etc.) then the lone pair is electron rich and so the -NH₂ group is **more basic**. If the -NH₂ group is joined to an **electron-withdrawing group** (the –I inductive effect of e.g. the delocalised electrons in an aromatic ring) then the lone pair is electron deficient and the -NH₂ group is **less basic**. The inductive effect explains the increase in basicity.

Ethylamine
is more basic than
ammonia
is more basic than
phenylamine.

TESTS

RECALL TEST

1 Write balanced equations for the reaction between ethanoic acid and:

   a NaOH(aq) _____

   b NaHCO$_3$(aq) _____

   c Na(s) _____

   d ethanol (with conc. H$_2$SO$_4$) _____

   e LiAlH$_4$ (dry ether) _____

2 State three tests for carboxylic acids:

3 State the reagents and conditions required to make ethyl propanoate.

4 Write a balanced equation for the acid hydrolysis of ethyl ethanoate.

5 State how may aspirin be prepared?

6 Write balanced equations for the following reactions:

   a (CH$_3$CO)$_2$O + H$_2$O →

   b (CH$_3$CO)$_2$O + CH$_3$CH$_2$OH →

   c CH$_3$CH$_2$Cl + NH$_3$ →

   d CH$_3$CH$_2$Cl + CH$_3$CH$_2$NH$_2$ →

   e CH$_3$NH$_2$ + HCl →

   f NH$_3$ + CH$_3$Br →

   g CH$_3$NH$_2$ + CH$_3$Br →

7 State the reagents and conditions required to convert the first substance into the second:

   a (CH$_3$CO)$_2$O into CH$_3$COOCH$_3$

   b (CH$_3$CO)$_2$O into CH$_3$CONHCH$_3$

Unit 3 TESTS

    c $CH_3NH_3^+$ into CH_3NH_2

    d CH_3COOH into $CH_3COO^-\ Na^+$

8 Give the reagents and conditions required to split the molecule $CH_3CH_2OCOCH_3$ in two ways:

CONCEPT TEST

1 This question concerns the following reactions:
$C_2H_4O_2$ (P) is converted by P_4O_{10} to $C_4H_6O_3$ (Q).
Q is converted by addition of ethanol to $C_4H_8O_2$ (R).
R is converted by aqueous NaOH to $C_2H_3O_2Na$ (S) and C_2H_6O (T).
P also may be converted by $LiAlH_4$ to T.
P reacts with NaOH(aq) to produce S.
T reacts with P to produce R.
CH_5N (U) reacts with P to produce C_3H_7NO (V).

  a Give the structures of P to V:

    P _____

    Q _____

    R _____

    S _____

    T _____

    U _____

    V _____

  b Using structural formulae, write equations for:

    i   the conversion of P to S

    ii the conversion of Q to R

    iii the conversion of U and P to V

    iv the conversion of R to S

  c State the conditions required to react:

    i   Q with T to make R

    ii P to make S

    iii R to make S and T

2 In this question you are given the name of the compounds and you must identify the reagents and conditions or draw the mechanism.

    a Aminomethane reacts with bromomethane. Draw the mechanism for this reaction.

    b Unsaturated fat and oils are in our diet.

       i Give a test to show that an oil is unsaturated

      ii Which type of unsaturated fat is thought to be harmful?

      iii Give a structural formula for the fatty acid named 10-hydroxydec-9-enoic acid which is found in Royal Jelly:

    c Give a test, and the expected result, that would distinguish between:

      i CH_3COOH and CH_3CH_2OH

      ii CH_3COOCH_3 and CH_3COOH.

3 Acetanilide (antifebrin) may be used in the manufacture of paracetamol. It has the formula, $C_6H_5NH(COCH_3)$.

    a Which two reagents could be used to prepare the solid Acetanilide. Note that ethanoic acid reacts too slowly to be of use.

    b Write a balanced equation for the reaction,

    c How could the solid be purified?

Unit 4

POLYMERS, OPTICAL ISOMERS, PROTEINS & AMINO ACIDS

AS organic chemistry is necessary for this topic.

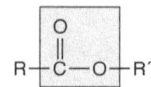

Fig. 4.1 Ester link

- There are two main groups of polymers: **addition polymers** and **condensation polymers**. Addition polymers were dealt with at AS level. Condensation polymers are the result of **condensation reactions**, often involving the ester or amide linkage. Ensure that you fully understand the four main condensation reactions (see unit 21: the reactions of carboxylic acids with alcohol and amines, also those of acid anhydrides with alcohol and amines).

- **Polyesters** are produced when **dicarboxylic acid** molecules **HOOC-[R]-COOH** condense with **diol** (dihydric alcohol) molecules **HO-[R']-OH**. R and R' are alkyl or aryl hydrocarbon residues. **Ester links** form between alternating acid and alcohol molecules (see Fig. 4.1).

  For example, benzene 1,4-dicarboxylic acid $HOOC(C_6H)COOH$ condenses with ethane-1,2-diol $HOCH_2CH_2OH$ to form **Terylene**, a common synthetic polyester fibre used in cloth and rope manufacture (see Fig. 4.2). A triol may be included to make cross links between polyester chains.

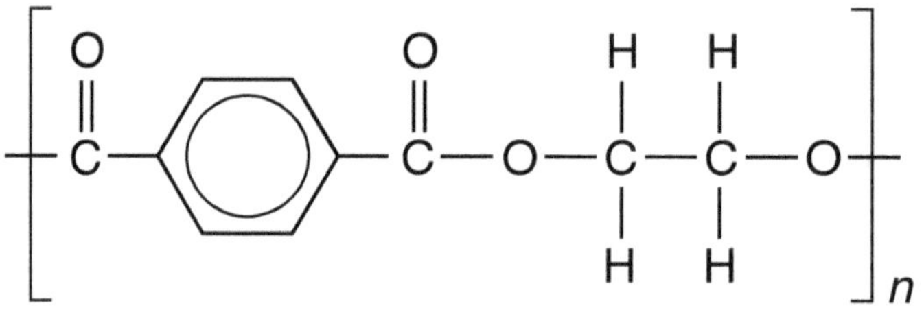

Fig. 4.2 Terylene repeating unit

Separate polyester chains are held together by **dipole–dipole** attraction due to the permanent **polarisation** in the **ester** groups, so polyesters have higher softening points than polyalkanes. These polymers are not crystalline, so they do not have sharp melting points.

- You must know one **biodegradable** polymer, such as **poly(lactic acid)**. It is made from a single monomer **2-hydroxypropanoic acid** $CH_3CH(OH)COOH$, which polymerises by forming ester links between the -OH and -COOH groups. The repeating unit is $[-OCH(CH_3)COO-]_n$.

- **Polyamides** such as **Nylon** are manufactured from a **diamine** $H_2N[R]NH_2$ and a **dicarboxylic acid** HOOC-[R']-COOH.

  For example, **Nylon 6,6** is made from 1,6-diaminohexane $H_2N(CH_2)_6NH_2$ and hexane-1,6-dioic acid $HOOC(CH_2)_4COOH$. Each monomer (made from benzene C_6H_6) contains 6 C atoms, hence the name Nylon 6,6. The repeating polymer unit is $[-HN(CH_2)_6NHCO(CH_2)_4COO-]_n$. **Amide links** couple the monomer units together (see Fig. 4.3).

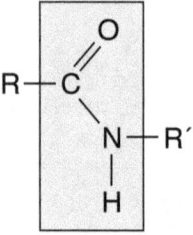

Fig. 4.3 Amide link

Nylon 6,10 is made from a diamine with six C atoms and a dicarboxylic acid with ten. These monomers can be derived from the castor oil plant.

A stronger type of Nylon is made by incorporating a benzene ring $-C_6H_4-$ into the structure. An example is **Kevlar**, which is used to make bullet-proof vests and puncture-proof canoes and tyres. It is made from 1,4-diaminobenzene $H_2NC_6H_4NH_2$ and benzene-1,4-dioic acid $HOOCC_6H_4COOH$. The repeating polymer unit is $[-HNC_6H_4NHCOC_6H_4COO-]_n$.

The Nylon polymer chains are strongly attracted to each other by **hydrogen** bonding between the -NH groups and the >C=O groups (see Fig. 4.4). Nylon is used as a fibre in clothes and carpets and in the heavy-duty ropes used to tie up ships. This tough plastic material is also used to make **hard-wearing** mechanical parts such as washing machine valves and food mixer gears.

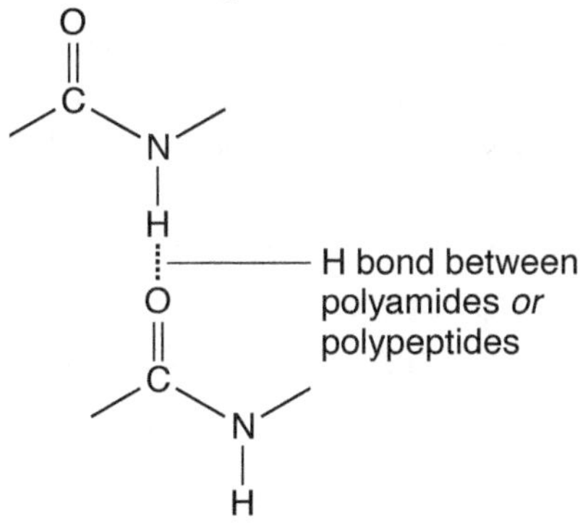

Fig. 4.4

- Because of their manufacture by **condensation reactions**, it should not surprise you to know that both polyesters and polyamides will be **hydrolysed** by prolonged exposure to aqueous acids and alkalis. Also these polymers may be photodegradable as the C=O absorbs ultra violet light.

- You must be able to recognise a polymer's **type** (addition, condensation, polyester, polyamide, etc.) from either the **repeating unit** (general formula) or a sample of the **chain**. Look for the type of link (amide, ester, or simple C-C) and then suggest the **monomers** used to form the links.

  When given monomer **names** or **structures**, you must be able to deduce the **type** of polymer and give the **repeating unit**.

- **Optical isomers** rotate the plane of **polarised light** in opposite directions. Two optical isomers (also called **enantiomers**) are mirror images of each other and cannot be superimposed on each other (see Fig. 4.1). Optical isomerism arises when a carbon atom (called a **chiral centre**) has four different atoms or groups attached to it.

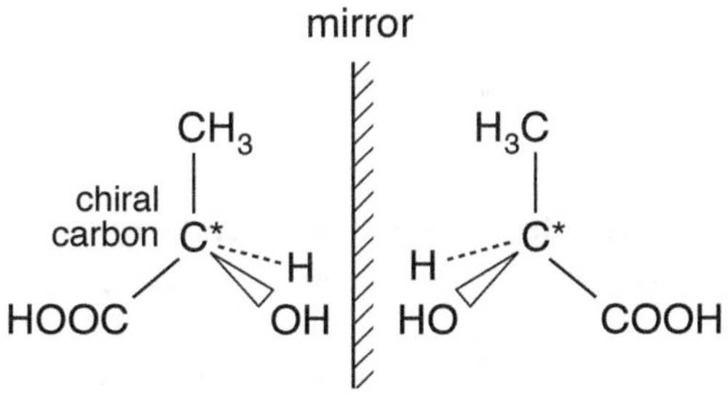

Fig. 4.1 Lactic acid shows optical isomerism.

Note that you must draw the isomers clearly in **three dimensions**. Take care to make the shape **tetrahedral** about the chiral centre.

Unit 4: POLYMERS, OPTICAL ISOMERS, PROTEINS & AMINO ACIDS

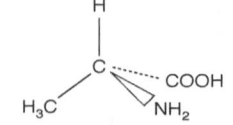

Fig. 4.6 Alanine, a simple amino acid

- **Amino acids** are substances that have both an **amine group** $-NH_2$ and a **carboxylic group** $-COOH$ (see Fig. 4.2). All naturally occurring amino acids have the general formula $H_2NCHRCOOH$. R may be a H atom or any one of a large variety of side chains. If R is not H then the C in the amino acid is **chiral**, so most amino acids have optical isomers.

- In the solid state and in aqueous solution, the acidic -COOH group donates a proton (H^+ ion) to the $-NH_2$ group, forming a double ion that is called a **zwitterion**.

  $H_2NCHRCOOH \rightleftharpoons {}^+H_3NCHRCOO^-$

  The zwitterion ion and has acidic and basic properties. An amino acid in solution exists as a zwitterion only at a particular pH value called the **isoelectric point**. Different amino acids, due to the different R groups, have different isoelectric points.

  When acid is added the zwitterion absorbs H^+ ions:

  ${}^+H_3NCHRCOO^- + H^+ \rightarrow {}^+H_3NCHRCOOH$

  When alkali is added the zwitterion donates H^+ ions:

  ${}^+H_3NCHRCOO^- + OH^- \rightarrow H_2NCHRCOO^- + H_2O$ (See Fig. 4.7)

Fig. 4.7

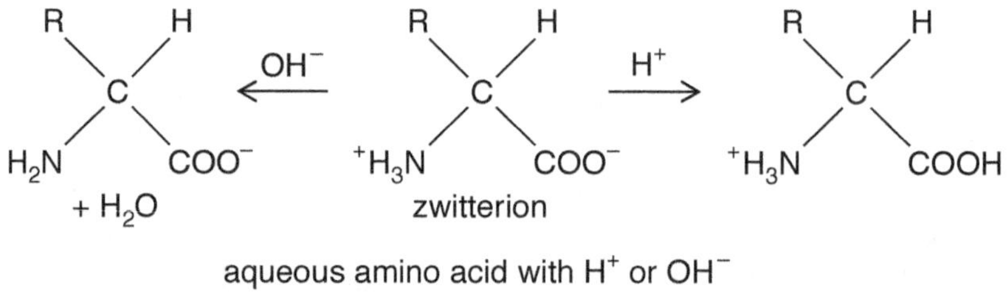

aqueous amino acid with H^+ or OH^-

- Amino acids condense together in chains to form **polypeptides**. The amino acids are joined by **peptide links**, which are the same as the **amide link** (see Fig. 4.8). Polypeptides join together to make **proteins**.

Fig. 4.8

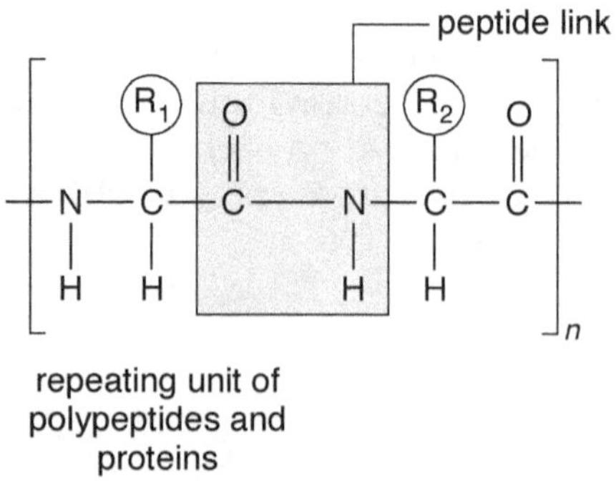

repeating unit of polypeptides and proteins

- 2-hydroxypropanoic acid (**lactic acid**) is found **naturally** in milk. It forms in your muscles when you run hard and makes them ache. Natural lactic acid always consists of just one optical isomer of 2-hydroxypropanoic acid because **enzymes** are **stereospecific** and only make one of the enantiomers.

  2-hydroxypropanoic acid can be chemically **synthesised** from ethanal without enzymes. The product is a **50–50 mixture** of the two optical isomers and is called a **racemic mixture**. The isomers have an equal and opposite effect on polarised light and so the plane of polarisation is **not rotated**.

TESTS

RECALL TEST

1 State the specific monomers required to make Terylene, a polyester.

2 Draw the repeating unit of
   a A polyester

   b Nylon 6,10

   c Polyhydroxybutanoic acid (PHB)

   d a polypeptide.

3 Give an example of a substance used to make cross links between the polyester chains in Terylene.

4 State the monomers required to make:

   a Nylon 6,6

   b Kevlar

   c protein

5 Give a large-scale use for each of these polymers:

   a Terylene

   b Nylon 6,6

   c Kevlar

Unit 4 TESTS

6 Boiling polyester and polyamides in aqueous alkali would .... the long-chain molecules. (Fill in the gap)

7 Give the formula of the species when:

   a amino acids are dissolved in water

   b acid is added to this solution

   c alkali is added to the original solution

8 State the type of polymer in each case (addition, polyester, polyamide, polypeptide):

   a $[-CH_2-CH(CH_3)-]_n$

   b $[-NHCH_2CONHCH_2CONHCH(CH_3)CO-]_n$

   c $[-O(CH_2)_2OCO(C_6H_4)CO-]_n$

   d $[-CO(C_6H_4)CONH(C_6H_4)NH-]_n$

   e $[-OCH(CH_3)CH_2CO-]n$

9 State what is meant by 'optical isomerism'.

10 Indicate which of the carbon atoms are chiral in this molecule:

$HOCH_2CHClCH_2CH(CH_3)COOH$

11 Explain how optical isomers are distinguished.

12 On a sheet of paper, draw the optical isomers of lactic acid (2-hydroxypropanoic acid).

CONCEPT TEST

1 Examine Fig. 4.10 below, which shows fragments of polymers.

polymer 1: $\cdots-\overset{O}{\underset{}{C}}-\underset{H}{N}-(CH_2)_6-\underset{H}{N}-\overset{O}{\underset{}{C}}-(CH_2)_4-\overset{O}{\underset{}{C}}-\underset{H}{N}-(CH_2)_6-\underset{H}{N}-\overset{O}{\underset{}{C}}-\cdots$

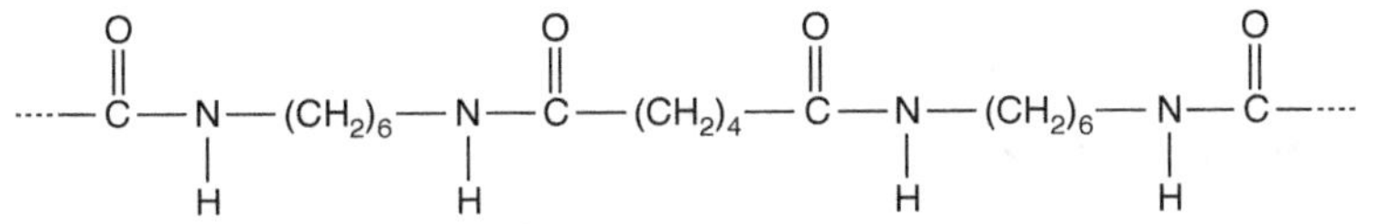

polymer 2: (peptide fragment with alternating amino acid residues having side chains CH_3, CH_3, CH_2CH_3)

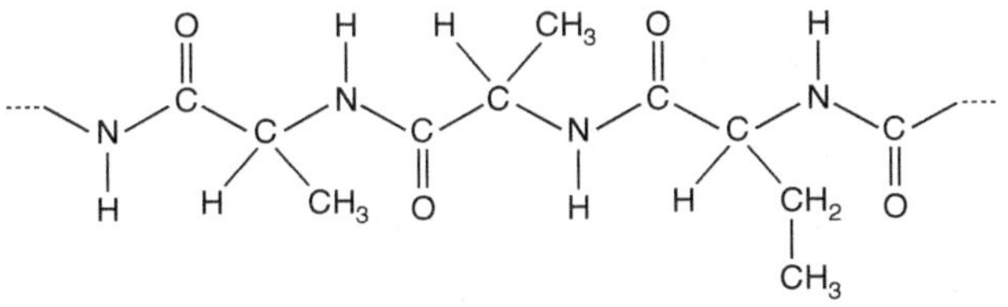

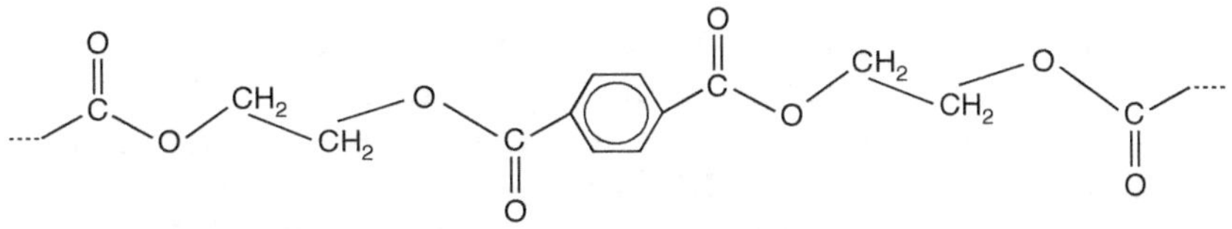

polymer 3: (polyester fragment containing $-O-CH_2-CH_2-O-$ groups linked to terephthalate units)

polymer 4: $\cdots-CH_2-CH_2-CH_2-CH_2-CH_2-CH_2-CH_2-CH_2-CH_2-CH_2-\cdots$

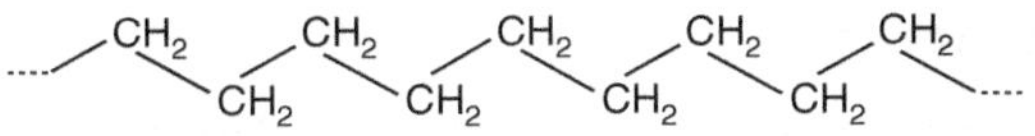

a Give the structural formula of the monomer(s) required for each polymer:

Polymer 1 _____

Polymer 2 _____

Polymer 3 _____

Polymer 4 _____

b State the type of polymer in each case:

Polymer 1 _____

Polymer 2 _____

Polymer 3 _____

Polymer 4 _____

c State the intermolecular forces in each case:

Polymer 1 _____

Polymer 2 _____

Polymer 3 _____

Polymer 4 _____

2 a Give the repeating unit for each polymer produced by these monomers:

  i $HOOC(C_6H_4)COOH$ with $H_2N(C_6H_4)NH_2$

  ii $H_2N(CH_2)_5COOH$

  iii $HO(CH_2)_2OH$ with $HOOC(C_6H_4)COOH$

b Give the type of the reaction required to make the polymer in part **a i**.

Unit 5

SYNTHESIS

AS organic chemistry is necessary for this topic.

- **Organic chemistry** builds up layer upon layer. If you do not recall a piece of the chemistry then other ideas that depend on it will be very difficult to grasp.

  Learn the material in this order:

  Functional groups

  Reaction types and **mechanisms** linked to functional groups

  The **reactions** of each functional group, including the AS reactions

  The **links** between the functional groups which produces a network

  The **chemical test** for each functional group

- Organic chemistry questions may be grouped into three types:

  You are given the **organic reactant, reagents and conditions**, and are required to state the **organic products**.
  You are given the **reactants and the products**, and are required to state the **reagents and conditions.**
  You are given the **products, reagents and condition**, and are required to state the **organic reactant**.

  Example: What is produced when propan-2-ol, $CH_3CH(OH)CH_3$, is heat under reflux with acidified potassium dichromate?

  $CH_3CH(OH)CH_3 + 2[O] \rightarrow$?

  Answer: potassium dichromate is an oxidising agent. Propan-2-ol is oxidised to propanone, CH_3COCH_3.

  Example: Which reagent and conditions are required to convert propan-2-ol, $CH_3CH(OH)CH_3$, to propanone, CH_3COCH_3?

  $CH_3CH(OH)CH_3$ + reagent? $\rightarrow CH_3COCH_3$

  Answer: The change from $CH_3CH(OH)CH_3$ to CH_3COCH_3 shows a gain of [O] and loss of [H] so oxidation has occurred so the reagent is acidified potassium dichromate.

  Example: Which reactant reacts with acidified potassium dichromate to make propanone, CH_3COCH_3?

  Reactant? + 2[O] $\rightarrow CH_3COCH_3$

  Answer: Potassium dichromate is an oxidising agent, and propan-2-ol will oxidise to form propanone.

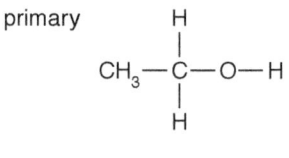

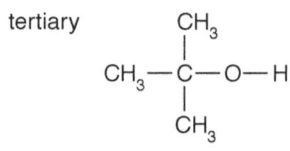

Fig. 5.1 Primary, secondary, and tertiary alcohols.

- **Draw** the graphic formula of the organic molecules you meet. Draw all the bonds. You will then see the functional groups more easily; you will notice which are primary, secondary, or tertiary; and it will stimulate you into seeing possible reactions.

- Functional groups may be **primary, secondary, or tertiary**. See Fig. 5.1. This may effect how a functional group reacts.

  Example: A primary alcohol may be oxidised to a carboxylic acid (when heated with $Cr_2O_7^{2-}/H^+$) or produce an aldehyde (when heated and distilled with $Cr_2O_7^{2-}/H^+$), and secondary alcohols produce a ketone (when heated with $Cr_2O_7^{2-}/H^+$), and tertiary alcohols are not oxidised when heated with $Cr_2O_7^{2-}/H^+$.

- One functional groups may make a **mixture of products**.

  Example: Using an acid catalyst, such as phosphoric acid, water is eliminated from butan-2-ol, $CH_3CH_2CH(OH)CH_3$. The -OH group is eliminated with another H atom which could be on the 1st or 3rd C atom, so the two products are but-1-ene and but-2-ene.

- Products may show **isomerisation**. They may be structural isomers, or stereo isomers (that is E-Z isomer or optical isomers). **Example:** Butan-2-ol shows optical isomerism. When water is eliminated the products but-1-ene and but-2-ene are structural isomers. But-2-ene shows E-Z isomerism.

- The molecules you meet may contain **several functional groups**, or contain two or more of the same functional group. Treat each functional group separately. Sometimes a reagent will react with one but not another functional group on the same molecule. Sometimes the reagent will react differently with the same functional group. Sometimes the same reagent will act differently with each functional group.

  Example: Consider $H_2NCHCHCOOCH_3$. What is produced when this molecule reacts with dilute hydrobromic acid, HBr(aq)? Drawing this molecule will show that there are three functional groups: $-NH_2$ an amine, a C=C which is an alkene, and -COOC- which is an ester link. In an acid-base reaction, the acid will donate a proton to the amine. By electrophilic addition the HBr will add to the C=C. The acid will hydrolyse the ester link. Thus the final products will be: $HOCH_3$ with a mixture of $^+H_3NCH_2CHBrCOOH$ and $^+H_3NCHBrCH_2COOH$.

- The questions may include one stage synthesis, two stages, or even **multistage synthesis**. This requires you to recall how one functional group maybe converted via intermediates and then into the final product. If you cannot see the answer straight away then consider what the starting molecule may be made into, and also what the final product may be made from. There may be a link, or you may be able to link two molecules from each group.

  Example: How is ethanoic acid prepared from ethene? Note that there is no direct route in one stage from ethene to ethanoic acid. Ethene may be made into ethane, a haloalkane, ethanol or polyethene. Ethanoic acid is made form ethanal and from ethanol. The intermediate is ethanol, so the stages in the preparation is: ethene → ethanol → ethanoic acid.

  Example: How is an azo dye produced from phenyl amine? You need to recall the many stages. See Fig. 5.2

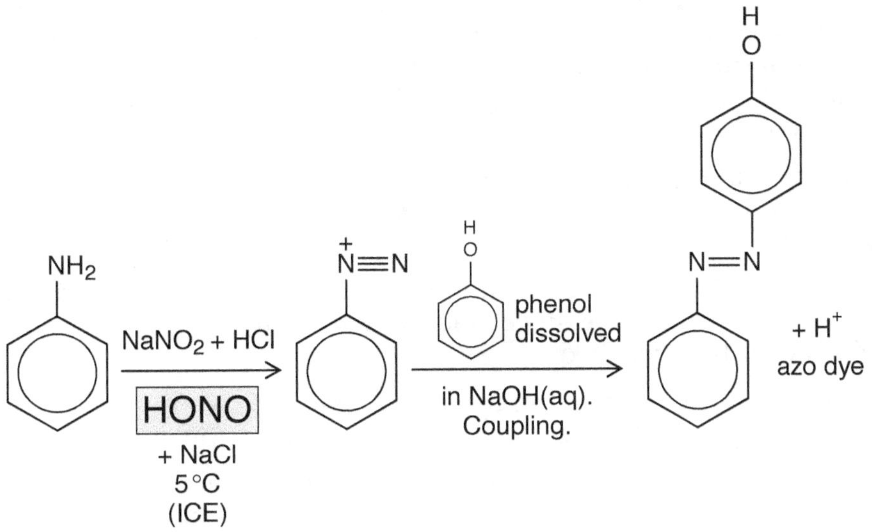

Fig. 5.2

Unit 5 SYNTHESIS

- **Pharmaceutical synthesis** produces most of the medicines and drugs we use. They may be referred to as **pharmaceuticals**.

    A **drug** is any substance that has a physiological effect. Potential new drugs are isolated by medical research. If it has a damaging effect on life it is called a **poison**. If it has a beneficial effect it is called a **medicine**. Some medicines are harmful if given to the wrong patients. For example, aspirin must not be given to children or people with stomach ulcers. The word 'pharmaceutical' is used to mean a drug or medicine.

- Drugs have a specific effect due to their **shape**. Hydrogen-bonding side groups such as -OH, NH_2, and -COOH help to make these substances **water-soluble.**

- **Chirality in pharmaceutical synthesis**: The synthesis of drugs and medicines often requires the production of a **single optical isomer**. One optical isomer may have beneficial effects where as the opposite one may have no effect or even be poisonous. **Example: Ibuprofen**, a pain killer and anti-inflammatory, has a chiral centre (see Fig. 5.3)

Fig. 5.3
$(CH_3)_2CHCH_2$— [benzene ring] —C*(CH_3)(H)—COOH

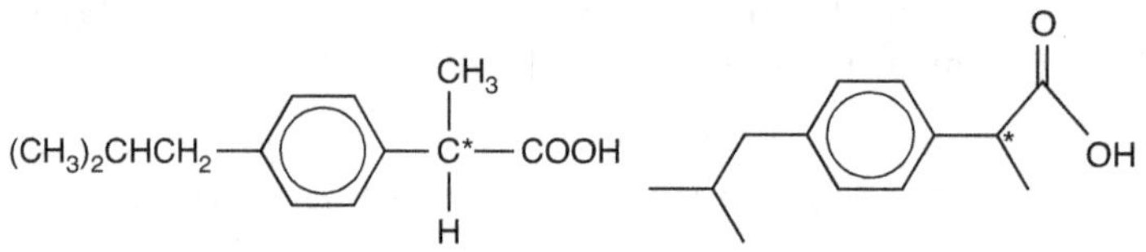

Ibuprofen (structural formula)

Ibuprofen (skeleton formula)

- When prepared synthetically in the laboratory molecules often contain a mixture of optical isomers. Whereas when produced by **enzymes** in living systems (e.g. bacteria or yeast) only one of the chiral molecules is produced.

    Chiral drugs and medicines are produced so as to **minimise side effects**, to **decrease cost**, and to **reduce risk** to companies from litigation.

    However the synthesis of a pharmaceutical that is a single optical isomer usually entails **increased costs** due to difficulty in separating the optical isomers, but it does reduce possible **side effects**, and usually improves the pharmacological activity (how effective the drug is).

    To produce a pharmaceutical with a single optical isomer is often carried out using purified **enzymes**, or **bacteria** (which contain enzymes), which promote **stereoselectivity**. Alternatively, **chiral catalysts** or chemical **chiral synthesis** is used. By using **natural chiral molecules**, such as L-amino acids or sugars, as starting materials usually only one chiral molecule is produced.

    Chemical chiral synthesis includes the of use **cyclic strained molecule**s, and the use of reagents fixed to a **polymer support** with reactants flowing over them.

    Supercritical CO_2 is used as a non-polar solvent. It is non-toxic and easily recycled. The product is easily extracted from the non-toxic supercritical CO_2 which is then recycled.

- The purity of drugs is often improved by **recrystallization**. Thin layer chromatography (**TLC**) is used as a simple check on purity and composition

- As already discussed: 2-hydroxypropanoic acid (**lactic acid**) is found **naturally** in milk. It forms in your muscles when you run hard and makes them ache. Natural lactic acid always consists of just one optical isomer of 2-hydroxypropanoic acid because **enzymes** are **stereospecific** and only make one of the enantiomers.

    2-hydroxypropanoic acid can be chemically **synthesised** from ethanal without enzymes. The product is a **50–50 mixture** of the two optical isomers and is called a **racemic mixture**. The isomers have an equal and opposite effect on polarised light and so the plane of polarisation is **not rotate**d.

TESTS

RECALL TEST

1. What is,

   a a pharmaceutical,

   b a chiral centre,

   c a tertiary haloalkane?

2. Explain why pharmaceutical companies prefer to produce a single optical isomer (3 reasons).

3. What is the disadvantage of producing a single optical isomer?

4. What is used to produce a single optical isomer? (Five points)

5. Give two examples of natural chiral molecules.

6. How may drugs be purified?

7. How is the purity of a drug tested?

8. State three methods used in chiral chemical synthesis.

Unit 6 — TESTS

CONCEPT TEST

1 Compound **P** has the formula:

   HOCH$_2$C(OH)(CH$_3$)CH(OH)CH$_3$

   a What type of stereoisomerism is shown by P? Draw the isomers.

   b State the structural formula of the products produced when P reacts with these reagents which are in excess:

   i Heated acidified potassium dichromate,

   ii Acidified potassium dichromate which is heated and distilled,

   iii Concentrated phosphoric acid,

   iv NaBH$_4$(aq)

   iv ethanoic acid (warmed with conc. H$_2$SO$_4$)

2 Compound **Q** is propan-1,2,3-triol.

   a Which organic compound would react with KOH(aq) to make Q?

   b What is produced if Q is reacted with excess ethanoic acid (warmed with conc. H$_2$SO$_4$)?

3 Ethanoic acid may be converted into methyl 2-aminoethanoate in three stages,

ethanoic acid → chloroethanoic acid → R → methyl 2-aminoethanoate

a The first stage produces chloroethanoic acid. State the reagents and conditions required.

b The second stage requires ammonia dissolved in ethanol to produce molecule R. Give the structural formula of R.

c State the reagents required to convert R into the final product methyl 2-aminoethanoate, $H_2NCH_2COOCH_3$.

d What type of compound is R?

e Does R show any stereoisomerism? Explain your reasoning.

f Give the formula of the compound produced when R is added to water.

g R will polymerise. Draw two repeating units of the polymer.

Unit 6

SPECTROSCOPY & CHROMATOGRAPHY

> AS organic chemistry is necessary for this topic.

> You must use all the evidence provided and not try to guess the answer from one clue.

> You may need other evidence before you can suggest a structural formula (so be patient).

- You are likely to meet questions in examinations that combine organic **analysis**, mass **spectroscopy**, infrared spectroscopy, and nuclear magnetic resonance (NMR) spectroscopy. Your aim is to use the evidence to **identify** an 'unknown compound' or suggest its structural formula.

- **Logical thinking** and **clear written presentation** will sort out which functional groups are present. You may also be able to calculate the relative formula mass (M_r).

- Used to separate out the components in a mixture, and to determine which substance is present, **chromatography** is a very important analytical tool, whether it is used to analyze samples from a crime scene, or used to determine the composition of an alien planetary atmosphere. Though the apparatus may be very different to paper chromatography, all forms of chromatography work in a similar way.

- In paper chromatography there is the paper (**the stationary phase**) and the solvent (**the mobile phase**), and the ratio (the R_f value) between the distance to the front of the solvent and the position of a coloured dot (the component) from the start position. So R_f is the distance moved by component divided by the distance moved by the solvent:

$$R_f = \frac{\text{migration distance of component}}{\text{migration distance of solvent front}}$$

> Make sure that you measure the migration distance from where the sample started and not from where the solvent started.

The distance moved by the coloured dot (the component) depends on how soluble it is in the solvent (the mobile phase) and attractive it is to the paper (the stationary phase).

- The **mobile phase** may be a gas or a liquid.

  The **stationary phase** maybe a solid, or a liquid supported on an inert solid.

- **Thin-layer chromatography, TLC**, is similar to paper chromatography. Instead of paper, the stationary phase is a thin layer of powdered solid, usually alumina (aluminium oxide) or silica (silicon dioxide) on a glass, plastic or aluminium plate. Here separation is due to the adsorption of the components by the solid phase. One advantage of TLC is that the powder that contains each component may be scraped off for further analysis.

- In **gas chromatography, GC**, really it is gas-liquid chromatography, the mobile phase is a gas (such as nitrogen or hydrogen) and the stationary phase is a liquid on the surface of a solid powder packed into a fine tube. Here separation is due to the solubility of the components in the liquid. One advantage of GC is that very small amounts may be analyzed.

  In GC, instead of the Rf value, the **retention time** is measured, which is the time measured from when the component is inserted to when it arrives at the end of the tube. Reference retention times are available for comparison for a particular gas and solid phase. However GC an unknown compound may have no reference retention times for comparison. The approximate relative proportions of each component is indicated by the size of the area under the peak.

- In **high pressure liquid chromatography, HPLC,** the stationary phase is held in a column and the mobile phase is a liquid which is forced through the stationary phase under high pressure. HPLC has the advantage of being fast.

- Usually GC and HPLC equipment are linked directly to a mass spectrometer (MS) and so called **GC-MS** or **HPLC-MS**. As each component arrives at the end (of the tube or column), it is passed directly into a mass spectrometer. The mass spectrum is compared with a computer spectral database so that it may be identified.

GC-MS and HPLC-MS are a much more powerful tool for analysis than chromatography alone. GC-MS is used in: **space probes** to analyse samples from other planets or moons; **forensics** to analyse samples from crime scenes; **airport security** to aid in the detection of explosives and drugs; and **environmental analysis**, for example to detect pesticides.

- **Mass spectra** were studied at AS. Remember that the **molecular ion** peak on the right of a spectrum gives the **relative molecular mass** of the compound. Other peaks will tell you the masses of the **fragments**.

- **Infrared spectroscopy**, studied at AS, may be linked to A2 spectroscopy.

- **Nuclear Magnetic Resonance**, NMR, is an effective analytical tool. NMR spectra give information about the **positions** of atoms in a molecule. The protons must be **unpaired** so hydrogen protons are investigated, or carbon-13 nuclei in C-13 NMR.

- **C-13 NMR** detects the environment of carbon atoms. As most carbon atoms are carbon-12, the signal is very weak and so it must be amplified. C-13 NMR spectra are easily read. Read the chemical shift of each peak and relate it to the table of chemical shifts given to you in the data booklet. Be careful to relate the spectra to any other evidence given. The sizes of the peaks (areas under the peaks) indicate the ratio of the carbon environments.

- In **proton-NMR** each hydrogen proton in a molecule absorbs energy from low energy **radio waves** at a particular frequency. Protons in different **environments** (positions) absorb at slightly different frequencies.

    Any solvent must be **hydrogen-free** so a solvent such as $CDCl_3$ is used. D indicates a deuterium atom. Atoms of deuterium (a hydrogen isotope) contain one neutron as well as one proton.

- Tetramethylsilane (**TMS**), $(CH_3)_4Si$, is used as a reference to calibrate NMR equipment because it gives a single peak on the spectrum. The methyl group hydrogen atoms all have the **same environment**. The TMS absorption is labelled **zero** and all other absorbancies are measured with reference to it.

    A simple **low-resolution** NMR spectrum shows vertical absorptions along the x-axis which is calibrated as a **chemical shift** in frequency relative to TMS = 0. You compare the spectrum against a table of chemical shifts for different proton environments (see Fig. 6.1).

- The **area** of each peak reflects the **relative number of protons** in that environment. **Example:** Ethanol CH_3CH_2OH has 3 H atoms in CH_3, 2 H atoms in CH_2, and 1 H atom in OH. Therefore the relative areas of the peaks CH_3: CH_2:OH will be 3:2:1 (see Figs 6.2 and 6.3).

> You don't need to explain how a nuclear magnetic resonance (**NMR**) spectrometer works.

> **Radio waves** are part of the electromagnetic spectrum that includes gamma rays, X-rays, visible light, and microwaves.

Type of proton	Chemical shift δ/p.p.m.
RCH_3	0.9
R_2CH_2	1.3
R_3CH	1.5
$R_2C=CH_2$	5
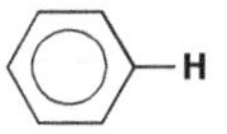 C₆H₅—H	7.3
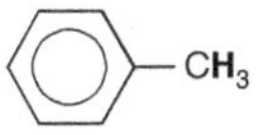 C₆H₅—CH_3	2.3
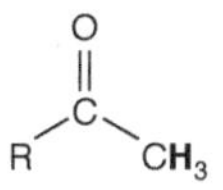 $R-C(=O)-CH_3$	2.3
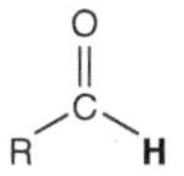 $R-C(=O)-H$	9.7

Fig. 6.1 Table of approximate chemical shifts and corresponding proton environments. Here R is an alkyl group.

Fig. 6.2

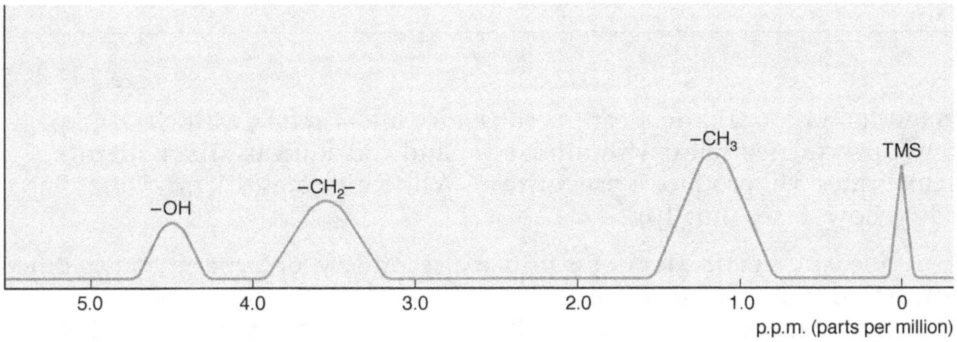

[NMR spectrum showing peaks at approximately 4.5 ppm (–OH), 3.5 ppm (–CH₂–), 1.0 ppm (–CH₃), and 0 ppm (TMS)]

One way of detecting -O-H or -N-H groups is by running the proton NMR, then add D_2O, deuterium oxide (heavy water). Any peak due to -O-H or -N-H group will disappear as the D will replace the H on the functional groups.

> The **data sheet** given to you in the exam shows the H-1 and C-13 NMR chemical shifts. You are allowed to have a copy outside exams.

Unit 6 — SPECTROSCOPY & CHROMATOGRAPHY

In the high-resolution ethanol spectrum (compare Figs 6.2 and 6.3), the CH_3 peak is split into 3 smaller peaks. The number of little peaks is equal to 1 plus the number of H atoms joined to the next atom. In this case the CH_2 group **splits** the CH_3 peak into 1 + 2 = 3 smaller peaks. The CH_3 group splits the CH_2 peak into 1 + 3 = 4 smaller peaks.

Fig. 6.3

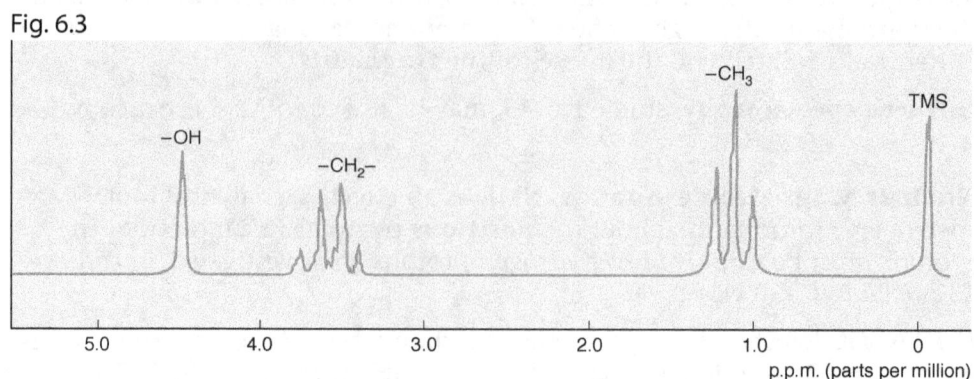

- You may need to identify **aromatic protons** (e.g. hydrogen atoms that are joined to a benzene ring) from chemical shift values. You will see a group of peaks with chemical shifts between 6.5 - 8.0. You will not be expected to analyse their splitting patterns.

- '**Magnetic resonance imaging**' (MRI) is used to obtain diagnostic information about the internal structures of our body. MRI is uses NMR technology.

> For each test you must know the **reagent**, any **observations**, and what a **positive result** indicates. Often one test will not tell you the functional group, but two together will.

- **Chemical tests** may be linked to A2 spectroscopy. They are much more specific, so are often carried out when there is some **clear idea** about the nature of the compound.

Common reagents	Positive test	Functional groups possible
Bromine solution	turns red/brown to colourless	alkene (>C=C<) or phenol (oily drops)
PCl_5(s)	vigorous reaction: white fumes of HCl(g)	-OH group (alcohol, carboxylic acid, water)
2,4-dinitro-phenylhydrazine	brightly coloured solid	ketone or aldehyde
Tollens' reagent	silver mirror or black solid	aldehyde present
$KHCO_3$(s)	fizzes: evolves CO_2	acid present (-COOH but not phenol

- To test for a **haloalkane**: heat the substance under reflux with NaOH(aq) to **hydrolyse**, neutralise with nitric acid, and add aqueous **silver nitrate**. Haloalkanes will produce a **precipitate**. White indicates -Cl, off-white -Br, pale yellow -I (see unit 10).

- To decide whether an **alcohol** is primary, secondary, or tertiary) add acidified **potassium dichromate**; colour change from orange to green indicates a **primary** or a **secondary** alcohol. Then test for an **aldehyde** using Tollen's reagent; a positive result indicates a **primary** alcohol.

> One ester smells of muscle rub: methyl 2-hydroxybenzoate ('Oil of Wintergreen').

- **Esters** generally smell **fruity**. Chemical **analysis** involves **hydrolysis** followed by identification of the **carboxylic acid** and the **alcohol**.

TESTS

RECALL TEST

1. Give the tests for each of these functional groups (the marks indicate the number of tests. Remember to state what occurs when the test is positive):

   a alkene

   b alcohol

   c carboxylic acid

   d ketone or aldehyde

   e aldehyde only

   f amine

   g a bromoalkane

   h phenol

2. Burning a sample in the fume cupboard may indicate something about a substance. State how the following burn differently:

   a a short-chain alcohol

   b an aromatic compound)

3. Molecules absorb infrared light because of

Unit 6 — TESTS

CONCEPT TEST

Fig. 6.4 Infrared absorption wave numbers

Bond	Wave number (cm^{-1})
C-H (alkane)	2850–2960
C-H (alkene)	3010–3900
C=C	1640–1680
C-O	1150–1310
C=O	1700–1740
O-H (hydrogen bonded)	3200–3550
O-H (not hydrogen bonded)	3590–3650
O-H (carboxylic acid)	2500–3300

Fig. 6.5 Table of approximate chemical shifts and corresponding proton environments. Here R is an alkyl group.

Type of proton	Chemical shift δ/p.p.m.
RCH$_3$	0.9
R$_2$CH$_2$	1.3
R$_3$CH	1.5
R$_2$C=CH$_2$	5
C$_6$H$_5$–H	7.3
C$_6$H$_5$–CH$_3$	2.3
R–CO–CH$_3$	2.3
R–CO–H	9.7

1 A mixture of oils were found in an ancient pottery container. A small sample was separated using steam distillation and gas chromatography. Three smelly components were P, cinnamaldehyde, C$_6$H$_5$CHCHCHO (which smells of cinnamon), Q, Oil of Wintergreen, HOC$_6$H$_4$COOCH$_3$ (which smells of muscle rub), and R, eugenol, CH$_2$CHCH$_2$C$_6$H$_3$(OH)OCH$_3$ (which smells of incense).

a State which compound(s) would react with the following reagents. State what would be observed and which functional group is detected.

  i Bromine water

  ii Silver nitrate dissolved in aqueous ammonia

  iii Aqueous potassium dichromate acidified with sulphuric acid

b See Fig. 6.6, the proton NMR of one of the compounds. Identify the compound and state the evidence that supports your answer.

Fig. 6.6

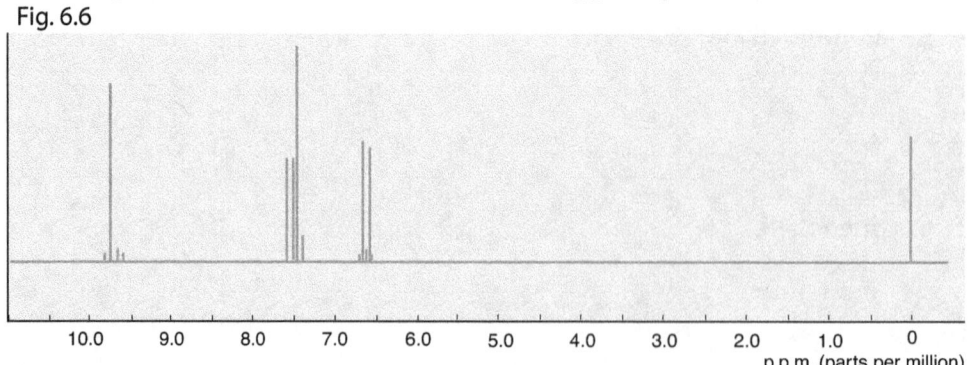

2 Compound X had a mass spectrum which produced peaks with mass/charge ratios of 17, 29, 76, 93, 105, and 122. The tallest peak was 76 and the shortest was 29. The following compounds could be X: CH_3COCH_2OH, CH_3CH_2COOH, $HOCH_2CH_2CH_3$, HOC_6H_4CHO, C_6H_5COOH, or $C_6H_5COCH_3$.

a Suggest the relative molecular mass of X.

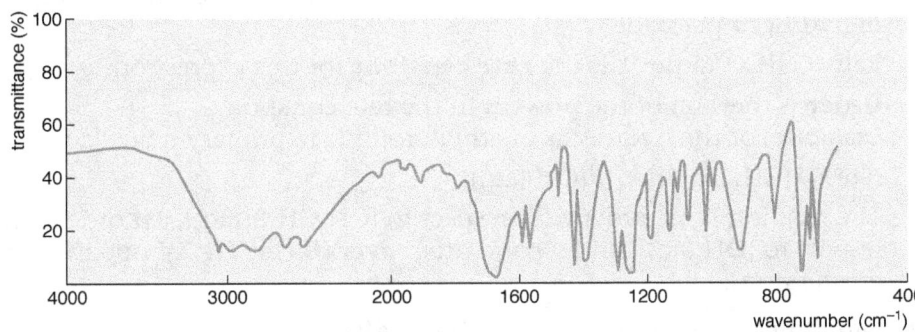

Fig. 6.7

b See Fig. 6.7, which shows the infrared spectrum of X, and, using the mass spectrum data, suggest what functional group(s) may be present in X, stating any evidence that supports your answer.

c Identify X.

d Give the C-13 chemical shift(s) you would expect for phenylethanone, $C_6H_4COCH_3$.

Unit 7

HOW FAST? RATES

You must understand the basic principles from the AS unit 2 before studying this one.

- You need to start with some **definitions**:

    Concentration has the units mol dm^{-3}. It is written with square brackets [ ].
    Example: For the decomposition of hydrogen peroxide

    $$2H_2O_2(aq) \rightarrow O_2(g) + 2H_2O(l)$$

    The concentration of hydrogen peroxide is written as $[H_2O_2(aq)]$.

    The rate equation links rate to the concentrations of the substances that control the reaction rate. **Example:** The rate equation for the decomposition of hydrogen peroxide is

    Rate = $k[H_2O_2(aq)]$ k is the **rate constant** for this expression.

    Order is the sum of the **powers** in the rate equation.
    Example: For the hydrolysis of ethyl bromide (a primary haloalkane),
    rate = $k[CH_3CH_2Br(aq)]^1[OH^-(aq)]^1$

    The reaction is 1st order with respect to $[CH_3CH_2Br(aq)]$, 1st order with respect to $[OH^-(aq)]$, and second order **overall** (see Fig 7.4 opposite for mechanism).

You will most likely find that all the reactions you meet in examinations will be zero, first, or second order with respect to each reactant.

- Order must be determined **experimentally**.

- Order suggests the number of particles involved in the rate determining step.

- The **slowest step** in a reaction pathway (mechanism) controls the overall reaction rate. This step is called the **rate determining step**.

    Example: For the iodination of propanone:

    $$CH_3COCH_3(aq) + I_2(aq) \rightarrow CH_3COCH_2I(aq) + HI(aq)$$

    Experiments show that $[CH_3COCH_3(aq)]$ and $[H^+(aq)]$ influence the rate, but not $[(I_2(aq)]$. Therefore, the rate equation is

    rate = $k[CH_3COCH_3(aq)]^1[H^+(aq)]^1$

    The rate equation shows that the CH_3COCH_3 and the H^+ are in the rate determining step. The balanced equation also shows I_2 as a reactant so another step in the mechanism must include the iodine.

Fig. 7.1

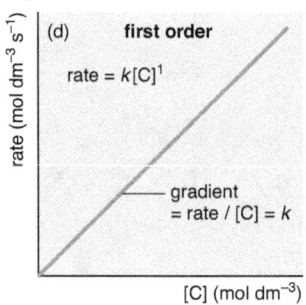

- Order is **first order** when the rate is proportional to the substance concentration. See Fig. 9.1.
    Example: In reactions where rate = $k[z]^1$
    Doubling the concentration of z doubles the rate.

Fig. 7.2

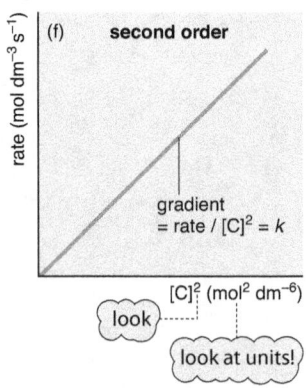

- Order is **second order** if the rate is proportional to the square of the substance concentration. See Fig. 9.2.
    Example: in reactions where rate = $k[z]^2$
    Doubling ($\times 2$) the concentration of z quadruples ($\times 4$) the rate; tripling ($\times 3$) the concentration of z multiplies the rate by 9.

- Order is **zero order** if the rate is independent of the substance concentration, i.e. varying the substance concentration has no effect on the rate. See Fig. 9.3.
    Rate = $k[z]^0$ i.e. rate = k because $[z]^0 = 1$.

Fig. 7.3

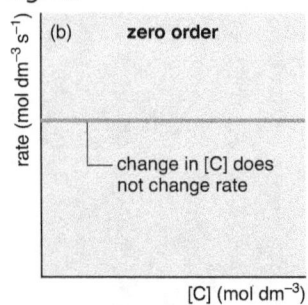

- Changes in **temperature** will change the rate constant and so the rate.

    If the temperature increases so the rate increases. This is due to any increase in temperature increasing the rate constant.

    In a similar way a decrease in temperature decreases the rate constant and so decreases the rate.

    Note that the only factor that influences the rate constant is temperature, for a particular reaction mechanism. An increase in concentration, for instance, does not influence the rate constant, but of course it may increase the rate.

- The **rate constant** is calculated by substituting values for concentration and for rate into the rate equation.

  Example: The rate equation is:
  rate = $k[CH_3COCH_3(aq)]^1[H^+(aq)]^1$
  If the rate is 0.02 mol dm^{-3} s^{-1}, $[CH_3COCH_3]$ = 0.1 mol dm^{-3}, $[H^+(aq)]$ = 0.01 mol dm^{-3}, $[I_2)]$ = 0.03 mol dm^{-3}, then re-arranging the rate equation:

  $k = \dfrac{\text{rate}}{[CH_3COCH_3(aq)]^1[H^+(aq)]^1} = 0.02/(0.1 \times 0.01) = 20$

  Units are determined by substituting the units into the rate equation and then cancelling down:

  $k = \dfrac{\text{mol dm}^{-3}\text{ s}^{-1}}{\text{mol dm}^{-3}\text{ mol dm}^{-3}} = \text{dm}^3\text{ mol}^{-1}\text{ s}^{-1}$

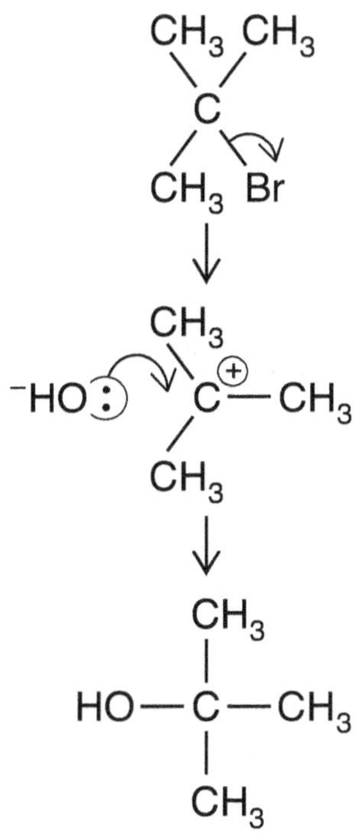

- During an exam, you may have to explain how to determine order by **experiment**. You must remember to state the three important practical points:

  (i) that the reaction is **repeated** several times under the **same conditions**;
  (ii) that the concentration of **one substance only** is changed at a time; and
  (iii) how the change in reactant concentration is **measured** (see below).

Property change	Instrument
Colour	Colorimeter
pH	pH meter
Electrical conductance	Conductivity meter
Plane of polarisation of light	Polarimeter
Gas volume	Gas syringe
Chemical change	Titration (in exams this one can involve complex explanations)

- **Half life** is the time taken for concentration of a reactant to fall by a half. See Fig. 7.5 to find out how to determine half life from a graph. Half life has a constant value for first-order reactions. See Fig. 7.6.

Fig. 7.5 How to calculate half life from a graph

To find out the half life, find out how long it takes for the concentration to half.

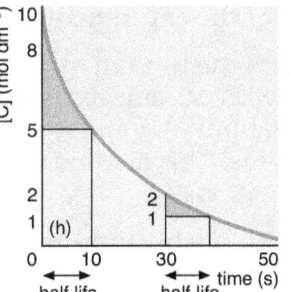

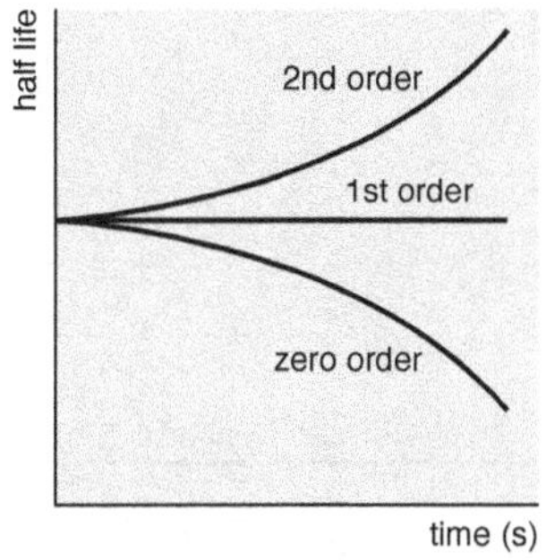

- **Rate** may be determined from a concentration / time graph. (See Fig. 7.7)

Fig. 7.7 How to calculate gradient from a graph

Draw a line that just touches the curve.
Read off the change in the y-axis and the x-axis.
Gradient = $\dfrac{\text{change in y}}{\text{change in x}}$

For a graph of concentration over time the gradient equals the rate.

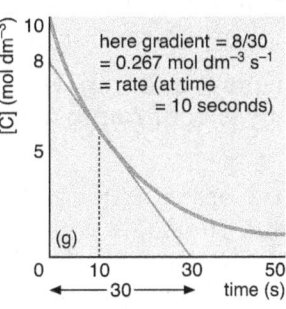

Fig. 7.4

Fig. 7.6

Unit 7

HOW FAST? RATES

- **Concentration / time graphs** are often used to determine order by either measuring the half life over time, or by calculating the change in rate over time, or by calculating the initial rate. See Figs. 7.5 to 7.10.

  Fig. 7.8 1st order graph of concentration against time

  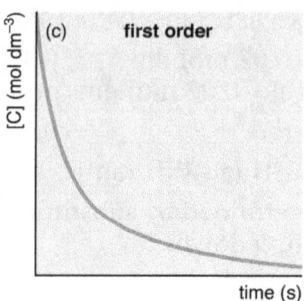

  Fig. 7.9 2nd order graph of concentration against time

  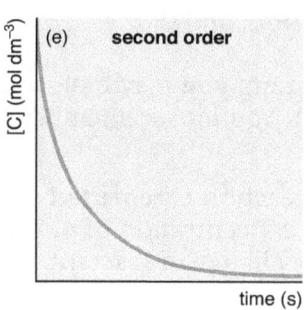

  Fig. 7.10 Zero order graph of concentration against time

  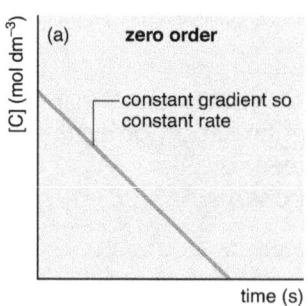

- **Initial rate** is the rate at the start of an experiment. To determine the initial rate draw a concentration vs time graph. Next measure the gradient at the start of the experiment. That initial gradient is the initial rate.

  Sometimes the time an experiment takes (for example to change colour) is recorded. The same experiment is run repeatedly using different concentrations. A graph may be drawn of the different concentrations against time. This method is not strictly accurate but does give good approximations.

- **Fig. 7.11** is a table showing results from four experiments. The rates shown are the initial rates of each experiment,

  The results of these experiments are not easy to interpret. However, you can compare the two experiments in which [q] is constant so you can see the effect of [p]. Experiments G and H show that, when [q] is constant, doubling [p] multiplies the rate by 4, showing that the reaction is **second order with respect to [p]** i.e. rate = $k'[p]^2$.

  Experiments G and E show that, when [q] is constant, tripling [p] multiplies the rate by 9, confirming that the reaction is **second order with respect to [p]**.

  To find the order of [q] you should look at experiments H and F, in which [p] is constant. Here, rate doubles when [q] doubles, showing that the reaction is **first order with respect to [q]**, i.e. rate = $k''[q]^1$. The reaction is third order overall i.e. **rate = $k[p]^2[q]^1$**.

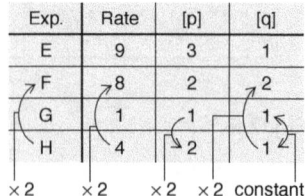

Fig. 7.11

TESTS

RECALL TEST

1. What is the rate equation? Give an example.

2. What is 'order'?

3. What are the units of k in a second order rate equation?

4. What are the units of k in a first order rate equation?

5. What are the units of k in a zero order rate equation?

6. What does the order suggest?

7. What is 'half life'?

8. State how these physical properties can be measured:

    colour

    pH

    electrical conductance

    polarised light

    gas volume

9. Finish the table of rate against [a] and [b], when [a] is first order and [b] is second order.

Rate	[a]	[b]
1	1	1
	2	1
	3	1
	1	2
	1	3

Unit 7 — TESTS

10 Draw a graph that summarises rate against half life for 1st, 2nd, and zero order reactants.

11 What does a straight-line graph of concentration against time suggest? What does the gradient tell you?

12 What does a straight-line graph of rate against concentration suggest? What does the gradient tell you?

13 What does a straight-line graph of rate against the square of concentration suggest? What does the gradient tell you?

14 Which graphs give straight lines with reagents that are

    a zero order?

    b first order?

    c second order?

CONCEPT TEST

1 The reaction of propanone (CH_3COCH_3) with iodine under acidic conditions was studied. The following data were collected:

$[CH_3COCH_3]$	$[I_2]$	$[H^+]$	Rate
0.1	0.1	0.1	2.0
0.2	0.1	0.1	4.0
0.1	0.2	0.1	2.0
0.1	0.1	0.2	2.0

    a Use the data to deduce the order of reaction with respect to these substances:

propanone

iodine

hydrogen ions

b What part did iodine play in the reaction?

c Write the rate equation for the reaction.

d State the minimum number of steps there must be in the mechanism.

e What must the role of the acid be in the reaction? Explain your answer.

f What is meant by the 'rate determining step'?

g Suggest which chemicals are involved in the rate determining step.

2 In the reaction between potassium permanganate and ethane-1,2-dioic acid, it is necessary to have acid conditions. This stoichiometric equation describes the reaction:

$2MnO_4^-(aq) + 5H_2C_2O_4 + 6H^+(aq) \rightarrow 2Mn^{2+}(aq) + 10CO_2(g) + 8H_2O(l)$

a From the data given, could you write a rate equation for this reaction? Explain your answer.

b How could the progress of the reaction be followed? State simply the method used, and which chemical concentration would be followed.

c State what is meant by the 'half life' of a reaction.

d What does the graph to the right suggest about the order with respect to X?

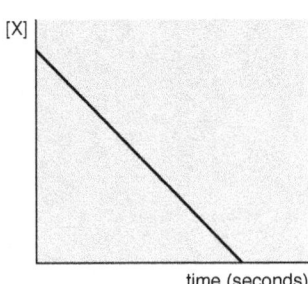

e In a rate equation it is thought that substance Y is first order. Suggest a graph that could be drawn to prove that rate is first order with respect to [Y].

Unit 8
HOW FAR? EQUILIBRIUM CALCULATIONS

You may need to review the AS unit 2 'How Far'.

- A **dynamic equilibrium** develops during all chemical reactions: **concentrations** of substances remain **constant** as reactants change into products (the **forward reaction**) and products revert to reactants (the **backward reaction** – or **reverse reaction**). At **equilibrium**, the **rates** of the forward and backward reactions are equal.

- **Le Chatelier's principle** states that if the conditions of a system at equilibrium are changed then the equilibrium position will **shift** to resist the change. You may need to recall this definition, but it is more important to understand how to apply the idea.

- K_c is the equilibrium constant for a reversible reaction written in terms of the **concentrations** of the reactants and products. The ratio involve the concentration of each substance raised to the same power as the number of moles shown in the balanced equation.

    For a general reaction: **reactants ⇌ products**,

    $$K_c = \frac{[\text{products}]}{[\text{reactants}]}$$

    Remember that square brackets [ ] stand for the concentration of a substance in mol dm^{-3}.

    Example: The production of ammonia NH_3 from hydrogen and nitrogen in the Haber process.

    $N_2(g) + 3H_2(g) \rightleftharpoons 2NH_3(g)$

    At equilibrium, $K_c = \dfrac{[NH_3(g)]^2_{eq}}{[H_2(g)]^3_{eq}\,[N_2(g)]_{eq}}$

    In this example, K_c has the units $\dfrac{(\text{mol dm}^{-3})^2}{(\text{mol dm}^{-3})^3\,(\text{mol dm}^{-3})}$

    i.e. mol^{-2} dm^6

    When the equilibrium constant does not have units, you should write (no units).

- You must remember that K_c is quoted at a **constant temperature**. Changing the temperature changes its value and the position of the equilibrium (so changes the proportions of reactants and products in the equilibrium mixture).

- Do **not include** water in the equilibrium expression for reactions that happen in **aqueous** solution; even when a reactant or a product, its concentration effectively remains constant.

    Include the concentration of water when it is a reactant or a product in **non-aqueous** reactions.

- It is important to recall that **only temperature** will effect the value of **Kc** (for a particular reaction).

  If the forward reaction is **exothermic** then an increase in the temperature will decrease the value of Kc, and so the equilibrium will shift to the left.

  If the forward reaction is **endothermic** then an increase in the temperature will increase the value of Kc, and so the equilibrium will shift to the right.

- Remember that changes in **concentration**, or the addition of a **catalyst**, have **no effect** on the value of Kc.
  When a concentration is changed then the equilibrium shifts to ensure that the value of Kc is correct.
  Adding a catalyst does not change the yield (the amount of product) but the yield is obtained more quickly.

  > The only factor to effect **Kc** is **temperature**.

- You may have to calculate K_c from given values of **equilibrium concentrations**.

  Consider this reaction:

  $CH_3COOH(l) + CH_3CH_2OH(l) \rightleftharpoons CH_3COOCH_2CH_3(l) + H_2O(l)$

  At equilibrium: $[CH_3COOH] = 0.01 \text{ mol dm}^{-3}$,
  $[CH_3CH_2OH] = 0.15 \text{ mol dm}^{-3}$
  $[CH_3COOCH_2CH_3] = 0.018 \text{ mol dm}^{-3}$,
  $[H_2O] = 0.02 \text{ mol dm}^{-3}$

  > Kc is **not** influenced by a change in concentration.

  $$K_c = \frac{[CH_3COOH] \times [CH_3CH_2OH]}{[CH_3COOCH_2CH_3] \times [H_2O]} = \frac{0.018 \times 0.02}{0.01 \times 0.15} = 0.24 \text{ no units}$$

  > Kc is **not** influenced by the addition of a catalyst.

- You may have to calculate K_c from given values of **initial** and **equilibrium** concentrations (see Fig. 10.1).

  $CO(g) + H_2(g) \rightleftharpoons C(s) + H_2O(g)$ at 600 °C (volume = 2.0 dm³)

  $K_c = \dfrac{[H_2O]}{[CO][H_2]}$

  Initially [CO] = 2.0 moles, [H₂] = 2.0 moles.
  At equilibrium [H₂O] = 0.40 moles.

	CO	H₂	H₂O	C
initial moles	2	2	0	0
eqm moles	2 − 0.4 = 1.6	2 − 0.4 = 1.6	0.4	0.4
eqm conc. mol dm⁻³	$\frac{1.6}{2.0} = 0.8$	$\frac{1.6}{2.0} = 0.8$	$\frac{0.4}{2.0} = 0.2$	$\frac{0.4}{2.0} = 0.2$

  $\therefore K_c = \dfrac{0.2}{0.8 \times 0.8} = 0.31 \text{ mol}^{-1} \text{dm}^3$

Fig. 8.1

Unit 8 — HOW FAR? EQUILIBRIUM CALCULATIONS

- You may have to work out equilibrium concentrations from given values of **initial concentrations** of reactants and K_c (see Fig. 8.2).

$A(aq) \rightleftharpoons B(aq)$

Initially there is 15 mol of A, and $K_c = 2.0$.

	[A]	[B]
initial	15	0
eqm	15 − x	x

$K_c = \dfrac{[B]}{[A]} = 2.0$

$2 = \dfrac{x}{15 - x}$ ∴ $30 - 2x = x$
∴ $3x = 30$
∴ $x = 10$ Therefore at eqm. $[A] = 15 - 10 = 5 \text{ mol dm}^{-3}$
   $[B] = 10 \text{ mol dm}^{-3}$

Fig. 8.2

- **Example:** The synthesis of ammonia.

$N_2(g) + 3H_2(g) \rightleftharpoons 2NH_3(g)$

Starting with 1.00 moles of N_2 and 3.00 moles of H, there were 0.20 moles of ammonia present in the equilibrium mixture at 800 K. Calculate K_c.

The remainder of the calculation is set out in Fig. 8.3.

	N_2	H_2	NH_3
initial moles	1	3	0
equilibrium moles	1 − 0.1 = 0.9	3 − 0.3 = 2.7	0.2

Fig. 8.3

$K_c = \dfrac{(0.2)^2}{0.9 \times (2.7)^3} = 2.2580 \times 10^{-3} = 2.25 \times 10^{-3}$

$\text{units} = \dfrac{(\text{mol dm}^{-3})^2}{\text{mol dm}^{-3} \times (\text{mol dm}^{-3})^3}$

$= \dfrac{1}{\text{mol}^2 \text{ dm}^{-6}}$

$= \text{mol}^{-2} \text{ dm}^6$

TESTS

RECALL TEST

1 What is meant by 'dynamic equilibrium'?

2 State Le Chatelier's principle.

3 alcohol + acid ⇌ ester + water (this reaction has ΔH ~ zero)

   State the how the value of Kc will change, and state which way the position of equilibrium will shift if the following conditions are applied to the reaction (state whether it shifts to the left, right, or does not change):

   a [alcohol] is increased

   b [ester] is increased

   c [acid] is decreased

   d temperature is increased

   e a catalyst is added)

4 $CaCO_3(s) \rightleftharpoons CaO(s) + CO_2(g)$ is an endothermic reaction, used to make basic calcium oxide, used to neutralise acid soils. State the how the value of Kc will change, and state which way the equilibrium will shift if the following conditions are applied to the reaction (state whether the reaction equilibrium shifts to the left, right, or does not change):

   a pressure is increased

   b temperature is decreased

   c pressure is decreased

5 $H_2(g) + I_2(g) \rightleftharpoons 2HI(g)$ is an exothermic reaction.

   State the how the value of Kc will change, and state whether the forward rate, backward rate, or yield increase under the following conditions:

   a increased temperature

   b increased pressure

   c addition of a catalyst

Unit 8 TESTS

6 The units of concentration are

$$+ CO_2(g)$$

7 a Give the expression for K_c for this reaction: $2SO_2(g) + O_2(g) \rightleftharpoons 2SO_3(g)$.

b What would the units be for this K_c?

8 The Wacker Process produces ethanal from ethene and oxygen:
$2CH_2CH_2(g) + O_2(g) \rightleftharpoons 2CH_3CHO(g)$.

a For this reaction write the equation for K_c.

b Write the units for this K_c.

9 The only thing to change the value of K_c or K_p is

CONCEPT TEST

1 The ester ethyl ethanoate may be produced by this reaction:

$$CH_3CH_2OH + CH_3COOH \rightleftharpoons CH_3COOCH_2CH_3 + H_2O$$

a Given that initially one mole each of CH_3CH_2OH and CH_3COOH were mixed together, and that at equilibrium 0.8 moles of CH_3COOH was found to be present, calculate the value of K_c.

b Often when making the ester in the laboratory, concentrated sulphuric acid is added. How does this increase the yield of ester?

c A catalyst may be added to the mixture. Explain the effect this would have on the yield at equilibrium.

2 This question concerns the efficiency and economics of the production of synthesis gas represented by this reaction:

$$CH_4(g) + H_2O(g) \rightleftharpoons CO(g) + 3H_2(g) \quad \Delta H = +523 \text{ kJ mol}^{-1}$$

a Describe and explain the effect of increasing the temperature on the yield.

b Give an expression for K_c for this reaction.

c What are the units of K_c for this reaction?

d State and explain the effect of increasing the temperature on the equilibrium constant K_c.

3 Hydrogen iodide will decompose when heated. This may be represented by $2HI(g) \rightleftharpoons H_2(g) + I_2(g)$.

a Give an expression for K_c for this reaction.

b What effect would doubling the all the concentration have on the amount decomposed? Explain your answer.

Unit 9: EQUILIBRIUM: ACIDS, BASES AND BUFFERS

You may need to review the AS unit 2 'How Far'.

- **The Brønsted–Lowry theory:** Acids are proton **donors** and bases are proton **acceptors**. Example:

  $HCl + H_2O \rightarrow H_3O^+ + Cl^-$
  Acid   Base   Acid   Base

  The HCl and Cl⁻ are a **conjugate pair**. Also H_2O and H_3O^+ are a conjugate pair.

  A **conjugate base** is made when its **conjugate acid** loses an H⁺ ion.

  Example:
  $NH_3 + H_2O \rightleftharpoons NH_4^+ + OH^-$
  Base   Acid   Acid   Base

  Here the NH_3 and NH_4^+ are paired. H_2O and OH^- are a conjugate pair. Also the conjugate acid of ammonia is the ammonium ion $NH_4^+(aq)$.

  Every acid has a corresponding **conjugate base** which is formed when the acid loses an proton. A conjugate base can accept a proton to regenerate the original undissociated acid.

  Every base has a corresponding **conjugate acid** which is formed when a base accepts a proton. A conjugate acid can lose a proton to regenerate the original undissociated base.

- **Strong** acids and bases are fully ionised in water. HCl(aq), H_2SO_4(aq), and HNO_3(aq) are strong acids; NaOH is a strong base.

- **Weak** acids and bases are partly ionised in water. H_2CO_3(aq), H_2SO_3(aq), and HNO_2(aq) are all weak acids; NH_3(aq) and CH_3NH_2(aq) are weak bases. Look for the $\rightleftharpoons$ sign.

  A **weak acid** such as ethanoic acid is partially dissociated:
  $CH_3COOH(aq) \rightleftharpoons CH_3CO_2^-(aq) + H^+(aq)$

  Here the conjugate base of ethanoic acid is the ethanoate ion $CH_3CO_2^-(aq)$.
  Remember that the alternative representation is:
  $CH_3COOH(aq) + H_2O(l) \rightleftharpoons CH_3CO_2^-(aq) + H_3O^+(aq)$

- **pH** is a measure of the concentration of H⁺(aq). $pH = -\log_{10}[H^+(aq)]$.

- **[H⁺(aq)]** may be calculated from **pH**: $[H^+(aq)] = 10^{-pH}$.

  Example: If a solution of hydrochloric acid has $[H^+(aq)] = 1 \times 10^{-5}$ mol dm⁻³ then pH = 5.0.

To convert [H⁺] to pH using a calculator: (i) type in the value of [H⁺(aq)]; (ii) press the log button; (iii) multiply by –1 (in your head or using the keypad).

- For any weak acid HA: $HA(aq) \rightleftharpoons H^+(aq) + A^-(aq)$

  This reversible reaction gives the equilibrium expression:

  $$K_a = \frac{[H^+(aq)][A^-(aq)]}{[HA(aq)]}$$

  K_a is the **acid dissociation constant**. The greater it is, the stronger the acid.

  Example: A solution is prepared by dissolving 0.01 moles of pure acid in water to make 1.0 dm³ of solution; $[H^+(aq)] = 0.001$ mol dm⁻³. Use the tabular method shown in Fig. 9.1 to calculate the K_a value.

Fig. 9.1 (concentrations given in mol dm⁻³)

	[HA]	[H⁺]	[A⁻]
initial	0.01	0	0
equilibrium	0.01 – 0.001 = 0.009	0.001	0.001

$$K_a = \frac{0.001 \times 0.001}{0.009}$$

$= 1.11 \times 10^{-4}$ mol dm⁻³

Example: A solution is prepared by dissolving 0.1 moles of benzoic acid in water to make 1.0 dm³ of solution.

Given that K_a for benzoic acid is 6.3×10^{-5} mol dm⁻³, use the method given in Fig. 9.2 to calculate the pH.

	[benzoic acid]	[H⁺]	[benzoate ions]
initial	0.1	0	0
equilibrium	0.1 – x = 0.1	x	x

Fig. 9.2 (concentrations given in mol dm⁻³)

so $6.3 \times 10^{-5} = \dfrac{x^2}{0.1}$ ∴ pH = 2.6

As x is very small $0.1 - x$ approximately equals 0.1. Do not make this approximation if you have the actual numeric values.

- K_a may be converted to pK_a to give numbers that are more convenient to manipulate and enable acid strengths to be compared more easily.

  pK_a = –log K_a

- The greater the value of K_a the stronger the acid, also the smaller the value of pKa. This is similar to [H⁺$_{(aq)}$] and pH. For when a solution is very acidic then the [H⁺$_{(aq)}$] is large and the pH is small.

- **Water dissociates** and sets up the following equilibrium:

  H₂O(l) ⇌ H⁺(aq) + OH⁻(aq)

  The dissociation is very slight so the concentration of water is unaffected.

  K_w = [H⁺(aq)] [OH⁻(aq)]

  K_w is the **ionic product of water** (the equilibrium constant for the ionization of water) and has the value 1.0×10^{-14} mol² dm⁻⁶ at room temperature.

- This is used to calculate the pH of a strong base. **Example:** Calculate the pH of 0.01 mol dm⁻³ solution of NaOH(aq).

  NaOH is a strong base so it is fully ionized, so [OH⁻(aq)] = [NaOH(aq)] = 0.01.
  K_w = [H⁺(aq)] [OH⁻(aq)] = 1.0×10^{-14} mol² dm⁻⁶.
  [H⁺(aq)] x 0.01 = 1.0×10^{-14} mol² dm⁻⁶, so [H⁺(aq)] = 1.0×10^{-12}.
  pH = $-\log_{10}$[H⁺$_{(aq)}$] = $-\log_{10}$ (1.0×10^{-12})
  So pH = 12.0

- **Buffer solutions** resist attempts to change their pH by the addition of (small amounts of) acid or base.
  Buffers are a **conjugate pair** formed from a **weak acid + conjugate base** pair or a **weak base + conjugate acid** pair.

- **Acid buffers** are made from a mixture of a **weak acid** and a **salt of a weak acid**. Example: Ethanoic acid with sodium ethanoate.

  The mixture produces a high concentration of both **CH₃COOH(aq)** and **CH₃CO₂⁻(aq)**.

  Added **H⁺(aq)** is removed from solution as it combines with **CH₃CO₂⁻(aq)**.
  CH₃COO⁻ + H⁺ → CH₃COOH

  Added **OH⁻(aq)** is removed from solution as it combines with H⁺(aq) from CH₃COOH(aq):
  CH₃COOH + OH⁻ → CH₃COO⁻ + H₂O

- **Blood plasma** is buffered by:

  H₂CO₃(aq) ⇌ HCO₃⁻(aq) + H⁺(aq)

  This maintains blood to a pH between 7.35 and 7.45. The weak acid is carbonic acid, H₂CO₃(aq), and the conjugate base is the hydrogencarbonate ion HCO₃⁻(aq).

Unit 9 — EQUILIBRIUM: ACIDS, BASES AND BUFFERS

- To **calculate acid buffer pH**, use the equation

$$pKa = pH - \log_{10} \frac{[A^-(aq)]}{[[HA(aq)]} \quad \text{i.e.} \quad pH = pK_a - \log_{10} \frac{[ACID]}{[[SALT]}$$

Example: Calculate the pH of a buffer solution in which $[CH_3COOH] = 0.1$ mol dm^{-3} and $[CH_3COO^-Na^+] = 0.05$ mol dm^{-3} (CH_3COOH pK_a = 4.76).
pH = 4.76 + $\log_{10}$ (0.05/0.1) = 4.76 + ($^-$0.30) = 4.46

- Note that, when an acid is **50% dissociated**, $[HA(aq)] = [A^-(aq)]$. Therefore, $K_a = [H^+(aq)]$, so **pK_a = pH**. (See indicators, below).

- **Indicators** are weak acids (or alkalis) which are coloured. When H$^+$ is lost (or gained) the colour changes. For an acidic indicator HIn:
$HIn(aq) \rightleftharpoons H^+(aq) + In^-(aq)$
The dissociation constant for an indicator is given the symbol K_{ind}.

$$K_{ind} = \frac{[H^+(aq)][In^-(aq)]}{[HIn]}$$

You must remember that when there are equal amounts of the two colours, then $[HIn(aq)] = [In^-]$ then $K_{ind} = [H^+(aq)]$, so pK_{ind} = pH. Therefore, the colour changes when **pH = pK_{ind}**. The colour change must occur within the abrupt change of pH (see the pH curves below). Indicators cannot show a distinct end point in weak acid – weak base titrations.

- The pH during a titration may be followed using a pH meter. The results may be presented as a pH curves (See 11.3).
The **equivalence point** is the point at which the acid and base are equimolar which is where the pH curve is vertical.
The **buffer range** is the part of the curve where the pH curve is approximately horizantal. **Example:** Study the pH curve to the top right in Fig. 9.3. The **buffer range** is roughly 5 cm^3 to 20 cm^3. The **pH range** of the buffer is 3-5 for the same curve. Note that when 12.5 cm^3 has been added then half the weak acid is neutralised by the strong alkali so then [acid] = [base], so pH = pKa.

Fig. 9.3 Using a **pH** meter to monitor changes in pH during a normal acid–base titration will produce the **titration curves** shown here.

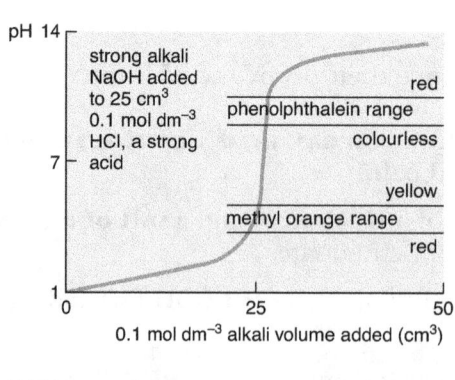
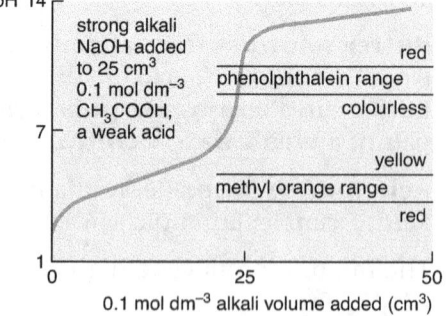
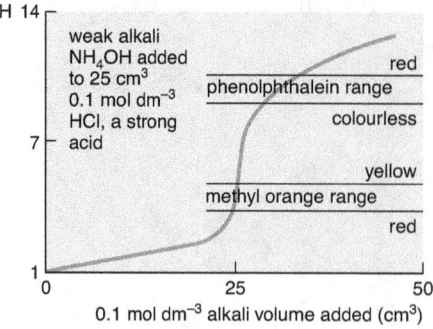
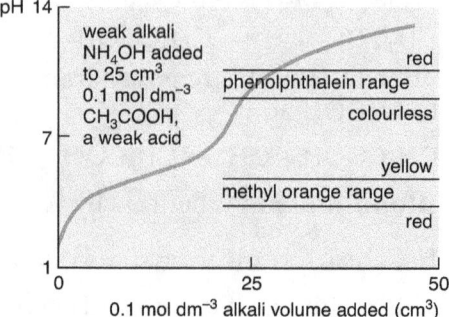

TESTS

RECALL TEST

1. A Brønsted–Lowry acid is
   and a base is

2. A strong acid is

3. Give equations for

   a pH

   b pK_a

   c $[H^+_{(aq)}]$ from pH

   d Ka from pKa

4. Find the pH of:

   a 0.05 mol dm^{-3} H$^+$ ions

   b 0.05 mol dm^{-3} HNO$_3$(aq)

   c 0.05 mol dm^{-3} H$_2$SO$_4$(aq)

   d 0.05 mol dm^{-3} CH$_3$COOH(aq) (when ethanoic acid $K_a = 5 \times 10^{-5}$ mol dm^{-3})

5. Write an equilibrium equation for the dissociation of methanoic acid (HCOOH).

6. Find the pH of 0.05 mol dm^{-3} NaOH(aq) (when $K_w = 10^{-14}$ mol^2 dm^{-6}).

7. What is a buffer solution?

8. What is the pH of a solution containing 0.05 mol dm^{-3} ethanoic acid and 0.05 mol dm^{-3} sodium ethanoate (when the pK_a of ethanoic acid is 4.76)?

Unit 9 TESTS

9 Calculate the pH of a solution of 0.05 mol dm^{-3} ethanoic acid and 0.1 mol dm^{-3} sodium ethanoate.

10 Draw a simple exam-quality titration curve for a strong acid with a weak base.

CONCEPT TEST

1 a Define

  i 'Brønsted–Lowry base'

  ii pH

  iii K_w

b Calculate the pH of a 0.01 mol dm^{-3} sulphuric acid solution.

c Ethanoic acid has pK_a = 4.76, and a concentration of 1.70 mol dm^{-3}. Calculate the pH.

d At 42 °C K_w is 3.0×10^{-14} mol^2 dm^{-6}. Calculate the pH of pure water at this temperature.

e At 42 °C, calculate the pH of a 0.001 mol dm^{-3} solution of NaOH.

2 Carbon dioxide dissolved in blood helps to buffer the blood pH.

$$HCO_3^-(aq) \rightleftharpoons H^+(aq) + CO_3^{2-}(aq)$$

a Use this equation to give the reactions that occur when acid and alkali are added.

acid added

alkali added

b 100 cm^3 of 0.2 mol dm^{-3} NaHCO$_3$ was mixed with 100 cm^3 of 0.1 mol dm^{-3} NaOH.

  i Calculate the concentration of CO$_3^{2-}$(aq) in the mixture.

  ii Calculate the concentration of HCO$_3^-$(aq) in the mixture.

  iii Calculate the pH of the mixture at 25 °C, given that K_a for the equilibrium above is 4.8 × 10^{-11} mol dm^{-3}.

Unit 10 — LATTICE ENTHALPY

AS topic enthalpy will help you with this topic.

- **Lattice enthalpy** is the standard enthalpy change when a solid ionic lattice is **formed** from its separate gaseous ions.
  Example: $Ca^{2+}(g) + 2Cl^-(g) \rightarrow CaCl_2(s) \quad \Delta H^\ominus_{latt} = -2237 \text{ kJ mol}^{-1}$

- Lattice enthalpy always has a negative sign.

- Ions with **higher charge** and smaller size (**smaller ionic radius**) produce a **greater** lattice enthalpy. Large values of lattice enthalpy indicate **high melting points**.
  Example: Ions in MgO are very small and doubly charged, so are strongly attracted to each other and **pack tightly** into their lattice. MgO has a very high melting point (2800 °C). It is used in refractory lining for furnaces.

The electron affinities of oxygen

1st EA -142 kJ mol^{-1}
$O(g) + e^- \longrightarrow O^-(g)$

2nd EA $+844 \text{ kJ mol}^{-1}$
$O^-(g) + e^- \longrightarrow O^{2-}(g)$

Fig. 10.1

- You must recall these definitions:
  1st electron affinity (EA): $X(g) + e^- \rightarrow X^-(g)$ The energy released when one mole of electrons is gained by one mole of gaseous atoms to form one mole of gaseous ions with a single negative charge.
  (**Slightly exothermic** as the nucleus in the neutral atom attracts the electron).
  2nd electron affinity: $X^-(g) + e^- \rightarrow X^{2-}(g)$ (**Greatly endothermic** as the negative ion repels the electron).
  Note the combined 1st and 2nd electron affinity is endothermic.
  1st and 2nd EA: $X(g) + 2e^- \rightarrow X^{2-}(g)$ (see Fig. 10.1.)

  Enthalpy of atomisation: The enthalpy change when one mole of gaseous atoms is formed from an element in its normal state under standard conditions. **Example:**
  $\frac{1}{2}H_2(g) \rightarrow H(g) \quad \Delta H^\ominus_a(H) = +218 \text{ kJ mol}^{-1}$

  Enthalpy of hydration: The enthalpy change when one mole of separate gaseous ions form hydrated ions (under standard conditions). **Example:**
  $Na^+(g) + \text{water} \rightarrow Na^+(aq) \quad \Delta H^\ominus_{hyd} = -406 \text{ kJ mol}^{-1}$

  Enthalpy of solution: The enthalpy change when one mole of a solid dissolves in a large excess of water (under standard conditions) to produce separate hydrated ions. **Example:**
  $NaOH(s) + \text{water} \rightarrow Na^+(aq) + OH^-(aq) \quad \Delta H^\ominus_{solu} = -42.7 \text{ kJ mol}^{-1}$

- You must know how to set out **Born–Haber cycles** (diagrams) that use Hess' law to calculate lattice enthalpies. You will find it useful to refer to the Born–Haber diagram for NaCl found in many text books.
  Example: The Born–Haber diagram to calculate the lattice formation enthalpy of $CaCl_2$ (see Fig. 10.2).

  1st ionisation energy of Ca = +590 $\Delta H_a(Cl) = +121$
  2nd ionisation energy of Ca = +1740 $\Delta H_a(Ca) = +193$ (data in kJ mol^{-1})
  electron affinity of Cl = −364 $\Delta H_f(CaCl_2) = -795$

  Comparing the two sides, from top of diagram to bottom:
  $^-(2 \times {}^+121) + {}^-(^+590 + 1150) + {}^-193 + {}^-795 = (2 \times {}^-728) + \text{lattice energy}$

  So the lattice enthalpy = $-2242 \text{ kJ mol}^{-1}$.

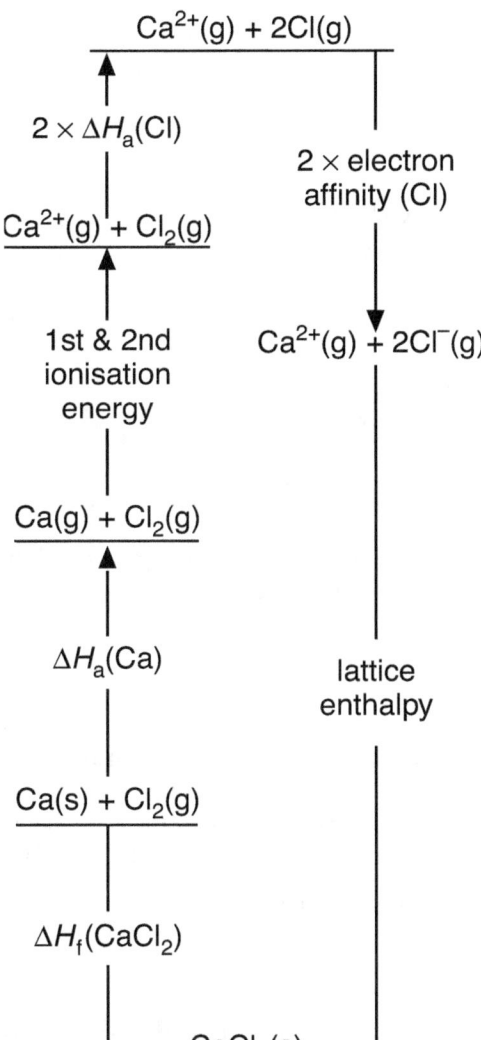

Fig. 10.2

- Lattice enthalpies **calculated** from Born–Haber cycles **differ** from those **predicted** by theoretical calculations based on a purely ionic model. See Fig. 10.3 for an example. These differences are evidence for the **covalent character** of many ionic solids.

	Experimental	Theoretical
AgF	960	920
AgCl	905	835
AgBr	890	815
AgI	890	780

Fig. 10.3 The lattice enthalpies (kJ mol^{-1}) of the silver halides

Unit 10 — LATTICE ENTHALPY

- Learn to combine lattice enthalpy, hydration enthalpy, and solution enthalpy into a calculation and into one diagram (see Fig. 10.4).

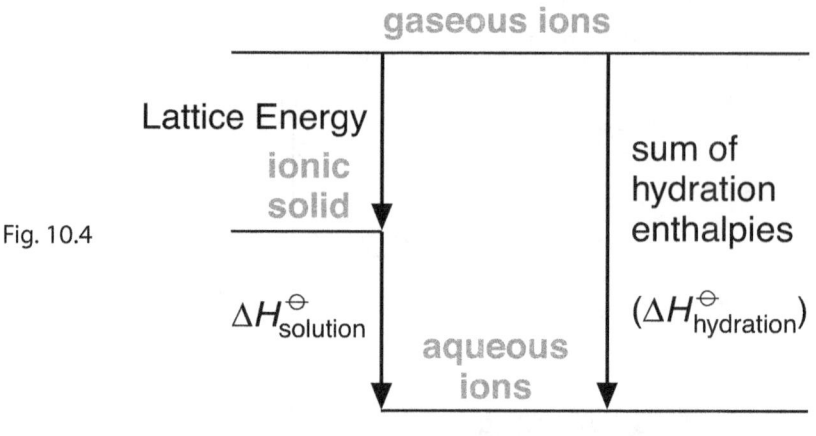

Fig. 10.4

Sum of hydration enthalpies = Lattice Energy + $\Delta H_{solution}$

- Ions with **higher charge** and smaller size (**smaller ionic radius**) produce a **greater** hydration enthalpy. Large values of hydration enthalpies may suggest that a substance is **soluble**, but if the lattice enthalpy is very much greater in magnitude then the substance will probably be insoluble.

  To correctly predict solubility the Gibbs free energy change, $\Delta G^\ominus$, must be calculated.

- Heat is evolved when solutions of an **acid** and an **alkali** are mixed. When dilute strong acid (e.g. aqueous HCl) neutralises dilute strong alkali (e.g. aqueous NaOH), the value of the **enthalpy of neutralisation** is always about -57 kJ mol^{-1}. Strong acids and strong alkalis are fully ionized in dilute solution, so the enthalpy change is always due to the reaction
  $H^+(aq) + OH^-(aq) \rightarrow H_2O(l)$
  All other ions are **spectator ions** and do not take part in the reaction.

- **Less heat** is evolved during neutralisation when either or both of the acid and alkali are **weak**. Note that the same number of moles of acid and alkali will react whether the acids and alkali are strong or weak. When a weak acid reacts with a strong alkali energy is required to break remove the H^+ ions from the acid molecule.

  Example: The reaction between dilute sodium hydroxide and ethanoic acid.
  $CH_3COOH(aq) + NaOH(aq) \rightarrow CH_3CO_2^-Na^+(aq) + H_2O(l)$ $\Delta H^\ominus_{neut} = -55.2$ kJ mol^{-1}
  During the reaction, energy is required to dissociate the weak acid into ions.
  $CH_3COOH(aq) \rightarrow CH_3CO_2^-(aq) + H^+(aq)$

TESTS

RECALL TEST

1 Define 'lattice enthalpy'.

2 Why does CaO have a greater lattice enthalpy than NaCl even though the ionic radii are similar?

3 Why does MgO have a greater lattice enthalpy than CaO?

4 Draw a Born–Haber diagram to show how the lattice enthalpy of sodium chloride could be determined.

5 Why is the theoretical lattice enthalpy of aluminium chloride so different to the calculated (experimental) value for lattice enthalpy?

Unit 10 — TESTS

6 Draw an energy cycle to show the connection between lattice enthalpy, hydration enthalpy, and solution enthalpy. What does the sum of the hydration enthalpies equal?

7 Using the terms 'lattice enthalpy' and 'hydration enthalpy', explain why magnesium hydroxide is insoluble and magnesium sulphate is soluble.

8 Explain the trend in lattice enthalpies of the group 2 oxides from MgO to BaO.

9 What is a strong acid?

11 Why do all reactions of strong acids with strong alkalis have approximately the same enthalpy of solution?

12 Why does the reaction of ethanoic acid and sodium hydroxide have a lower enthalpy of neutralisation?

CONCEPT TEST

$kJ\,mol^{-1}$

$\Delta H_{formation}(CaO) = -635$

$\Delta H_{bond\,enthalpy}$ of $O=O$ $= +496$

$\Delta H_a(Ca) = +193$

1st electron affinity of O $= -142$

1st ionisation energy of Ca $= +590$

2nd electron affinity of O $= +844$

2nd ionisation energy of Ca $= +115$

1 a Draw a Born–Haber diagram and calculate the lattice enthalpy of calcium oxide. The values are to the left.

b Consider a similar diagram for CaCl₃. Why is CaCl₃ never made?

c Explain why the theoretical lattice enthalpy of AgI is so different from the experimental lattice enthalpy calculated from a Born–Haber diagram.

d Sodium chloride is well known to be soluble.

    i Some relevant information is shown right. Calculate the $\Delta H_{solution}$ of sodium chloride.

kJ mol⁻¹

$\Delta H_{hydration}$ of sodium ions = –406

$\Delta H_{hydration}$ of chloride ions = –364

lattice enthalpy of sodium chloride = +771

    ii Explain why sodium chloride dissolves in terms of the particles, the forces between particles, and the arrangements and kinetic energies of the particles.

2 When aqueous KOH reacts with aqueous HCl, both substances are fully ionised. When hydrogen cyanide reacts with KOH the HCN is hardly ionised. The reactions may be expressed as follows:

Reaction I: KOH(aq) + HCl(aq) → KCl(aq) + H₂O(l) $\Delta H_n = -57.2$ kJ mol⁻¹

Reaction II: KOH(aq) + HCN(aq) → KCN(aq) + H₂O(l) $\Delta H_n = -11.6$ kJ mol⁻¹

Reaction III: HCN(aq) → H⁺(aq) + CN⁻(aq)

Assuming aqueous KCl and aqueous KOH are fully ionised, calculate a value for reaction III.

Unit 11: ENTHALPY AND ENTROPY

AS topic enthalpy will help you with this topic.

HCl(aq), H_2SO_4(aq), and HNO_3(aq) are strong acids; NaOH is a strong base.

- **Entropy** is a measure of **disorder**. Entropy (symbol S) increases when disorder increases.

  Put milk into coffee, shuffle a pack of cards, throw your clothes randomly around your room. In all these situations the objects will become more disordered. You would not expect the coffee and milk to separate, or the pack of cards to be sorted after shuffling, or the room to be tidy.

  Everyday examples of entropy increasing also include a cup of hot coffee cooling down (energy and water molecules spreading into the atmosphere), ice melting, and gases mixing. Burning is an exothermic reaction which includes an increase in entropy. It may seem strange that you can measure disorder, but you can, and it enables you to decide whether a reaction is **feasible**.

- The idea is that the most likely thing to happen is disorder. As there is such a large number of particles involved in chemical reactions the probability of something becoming organised by itself is extremely small. For example, if two coloured gases were mixed and left then they would not unmix, even if they were left for the lifetime of the universe.

- Things do become organised (eg. crystals do grow from solution), but only if something else becomes more disordered. Heating things up makes them more stirred up, disordered. The heat transferred to the surroundings enables crystals to form.

- Remember; The most likely thing is disorder. Like untidy rooms.

- A great **increase** in entropy occurs when:
  A **solid** changes into a **gas**,
  A **solid** dissolves to form a **solution**,
  A reaction shows an **increase in gas molecules.**

- $\Delta H^\ominus$ is the standard enthalpy change that accompanies a reaction. **Enthalpy** and **entropy** are **connected** by the equation:

  $\Delta G^\ominus = \Delta H^\ominus - T\Delta S^\ominus$

  $\Delta G^\ominus$ is the change in **Gibbs energy**, that is, energy that is **free** to do work.
  T is temperature in Kelvin.
  This equation is derived on page 64.

- Feasible, **spontaneous reactions** (under standard conditions) are indicated when $\Delta G^\ominus$ has a **negative value**. Equilibria lie further to the right with increasingly negative values.

  *For a feasible reaction $\Delta G^\ominus$ must be **negative***

  $\Delta G^\ominus$ will have a **negative value** either
  (i) when $\Delta H^\ominus$ **is negative** (the reaction is exothermic which usually means $T\Delta S^\ominus$ is **very small**), or
  (ii) when $\Delta H^\ominus$ **is positive** (the reaction is endothermic) and the $T\Delta S^\ominus$ term is a **large positive number**, e.g. the reaction between $KHCO_3$ and acid is an endothermic change; it is spontaneous, because it gives off a gas, which involves a large increase in entropy.

  Generally then, for a reaction to be feasible, and so $\Delta G^\ominus$ to be negative, then $\Delta H^\ominus$ must be less than $T\Delta S^\ominus$.

  For a feasible spontaneous change: $T\Delta S^\ominus > \Delta H^\ominus$

- Sometimes, reactions **do not occur** when $\Delta G^\ominus$ suggests they should. In these cases, you should state that there must be a **high activation energy**, so the rate is low. The reactants are **kinetically stable** due to activation energy (E_a) but are **thermodynamically unstable** due to $\Delta G^\ominus$.

- Sometimes the $\Delta G^\ominus$ term predicts that a reaction is not possible, but it occurs spontaneously **in practice**. In these cases, you should state that **non-standard** conditions are being used. You should work out which conditions are non-standard.

- Often you may rightly suggest a reaction is **feasible** because $\Delta H^\ominus$ **is negative**. You may then state that the reaction may **not be feasible** due to a **large negative value of** $\Delta S^\ominus$, corresponding to an increase in order.

> Remember that temperature must be in Kelvin, K.
>
> To change degrees Celsius, °C, into K, just add 273.15.
>
> Kelvin = °C + 273.15

- **Calculations** involving entropy and Gibbs energy may look difficult but they are fairly straightforward. Here are some possible types of calculation you might meet:

  (a) The examiner may give you values of $\Delta S^\ominus$, T, and $\Delta H^\ominus$ and you have to calculate $\Delta G^\ominus$ (use $\Delta G^\ominus = \Delta H^\ominus - T\Delta S^\ominus$). **Example:** Calculate the value of $\Delta G^\ominus$ for the combustion of carbon (graphite) given that:
  $\Delta H^\ominus_{(c)} = -393.5$ kJ mol^{-1} = $-393.5 \times 1000 = -393\,500$ J mol^{-1}
  T = 25 °C = 298 K
  $\Delta S^\ominus_{(c)} = 5$ J K^{-1} mol^{-1}
  $\Delta G^\ominus = \Delta H^\ominus - T\Delta S^\ominus$, so $\Delta G^\ominus$
  = -393 500 - (298 x 214) = -394990.0
  = -395 kJ mol^{-1} (to 3 significant figures).

> Take care to use the correct units.
> You may be given an enthalpy, ΔH, in kJ mol^{-1}, and a ΔS in J K^{-1} mol^{-1}.
>
> Convert the kJ into J by multiplying by 1000.

  (b) Conversely, you may be given values for $\Delta G^\ominus$, T, $\Delta H^\ominus$ and have to calculate (use $\Delta G^\ominus = \Delta H^\ominus - T\Delta S^\ominus$).
  Example: Calculate the value of $\Delta S^\ominus$ for the combustion of hydrogen given:
  $H_2(g) + \tfrac{1}{2}O_2(g) \rightarrow H_2O(l)$
  $\Delta H^\ominus_{(c)} = -286$ kJ mol^{-1} = $-286 \times 1000 = -286\,000$ J mol^{-1}
  T = 25 °C = 298 K
  $\Delta G^\ominus_{(c)} = -273$ kJ mol^{-1} = $-273 \times 1000 = -273\,000$ J mol^{-1}
  Rearranging $\Delta G^\ominus = \Delta H^\ominus - T\Delta S^\ominus$,
  $\Delta S^\ominus = (\Delta H^\ominus - \Delta G^\ominus)/T$ = (-286 000 - ⁻273 000)/298 = (-286 000 + 273 000)/298 = -43.624 = -43.6 J K^{-1} mol^{-1}

  (c) You may be given values of $\Delta S^\ominus$ for several different reactions and have to calculate $\Delta S^\ominus$ for another reaction. You simply construct an entropy cycle diagram involving $\Delta S^\ominus$ terms, in the same way that you would construct a Hess cycle using ΔH terms.

  (d) You may have to calculate the temperature when ΔG is zero which is the temperature when the reaction changes from going one way to another.

Note these examples:

- Burning is an exothermic reaction which includes an increase in entropy.
- Life is an example of a reaction in which there is a decrease in entropy (the molecules in life become more ordered) due to the heat given off. Overall there is an increase in order.

> **Example:** Water does not freeze (an **exothermic** change) at room temperature because it involves making ordered ice crystals which is accompanied by a large **entropy decrease**.

- Melting, evaporation and dissolving potassium aluminium sulphate are examples of changes which are endothermic but occur due to an increase in entropy, so the particles become more disordered.
- Potassium or sodium hydrogen carbonate (KHCO3 or NaHCO3) with acid are examples of reactions which are endothermic, but which occur due to an increase in disorder, (entropy)
- There are no examples of endothermic changes which also increase order. Ash and air will not spontaneously form paper.

Unit 11 — ENTHALPY AND ENTROPY

- This is how to derive the above equation $\Delta G^\ominus = \Delta H^\ominus - T\Delta S^\ominus$:

The reactants and products involved in a chemical reaction are called a **system**. The rest of the universe is called the **surroundings**. It is possible to calculate the standard **entropy change** $\Delta S^\ominus$ of a system during a reaction, as reactants change into products. See Fig. 11.1

Fig. 11.1

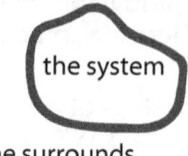

the surrounds

Disorder always increases, so the total entropy change of the universe, that is called $\Delta S^\ominus_{total}$, must always be positive if a reaction is feasible.

Any system, like a group of molecules, could be thought of as the system (the chemicals) and the surrounds (the total universe less the chemicals), so
Equation 1: $\Delta S^\ominus_{total} = \Delta S^\ominus_{surrounds} + \Delta S^\ominus_{system}$.

Now $\Delta S^\ominus_{surrounds} = \Delta H^\ominus / T$,

which means the entropy change of the surrounds is equal to the heat produced by the reaction divided by temperature. This may seem odd, but at a low temperature, say near zero K, an increasing temperature by 1 K (1°C) would dramatically increase entropy, whereas a 1 K increase at 1000 K would have only a slight effect. The negative sign is there because exothermic changes have a negative sign.

Substituting this into Equation 1 produces:

$\Delta S^\ominus_{total} = - \Delta H^\ominus_{system}/T + \Delta S^\ominus_{system}$.

Multiplying by temperature and changing the signs produces:

$-T\Delta S^\ominus_{total} = -T(-\Delta H^\ominus_{system})/T + -T\Delta S^\ominus_{system}$.

which after cancelling and rearranging:

$-T\Delta S^\ominus_{total} = \Delta H^\ominus_{system} - T\Delta S^\ominus_{system}$

and as $\Delta G^\ominus = -T\Delta S^\ominus_{total}$ we end up with the equation:

$\Delta G^\ominus = \Delta H^\ominus_{system} - T\Delta S^\ominus_{system}$ we end up with the equation:

As $\Delta S^\ominus_{total}$ must always be positive, so $-T\Delta S^\ominus_{total}$ must always be negative. Any reaction that is feasible must have a negative $\Delta G^\ominus$.

TESTS

RECALL TEST

1. What is meant by entropy?

2. What symbol is used to denote entropy?

3. State whether the follow show an increase or decrease or no change in entropy:

    a The melting of solid wax,

    b Iodine solid subliming to form iodine gas,

    c Sodium chloride crystals dissolving in water,

    d Water vapour (a gas) condensing to form ice in a freezer,

    e The formation of ammonia: $N_2(g) + 3H_2(g) \rightarrow 2NH_3(g)$

    f The thermal decomposition of limestone:
    $CaCO_3(s) \rightarrow CaO(g) + CO_2(s)$

4. For a reaction to be spontaneous (feasible) what must $\Delta G^\ominus$ be?

5. Explain why some endothermic reactions are feasible.

6. Explain why some reactions spontaneously produce a more ordered set of molecules.

7. The $\Delta G^\ominus$ value for the reaction of methane with oxygen indicates that a reaction is feasible but it does occur at room temperature. Explain this.

Unit 11 — TESTS

CONCEPT TEST

1. The production of sulfuric acid requires many stages.

   a First sulfur is burnt in oxygen to make sulfur dioxide:
   $$S(s) + O_2(g) \rightarrow SO_2(g)$$
   This reaction is spontaneous at 900 K.
   Predict the sign of ΔG for the reaction. Give your reasoning.

   b To make sulfuric acid sulfur dioxide is reacted with oxygen to make sulfur trioxide:
   $$SO_2(g) + \tfrac{1}{2}O_2(g) \rightarrow SO_3(g)$$

   i In terms of the arrangement of the particles, do the particles in this reaction show an increase or decrease in entropy? Explain your answer.

   ii At 450°C:

   $\Delta H^\ominus$ of reaction = -98 kJ mol^{-1}

   $\Delta G^\ominus$ of reaction = -70 kJ mol^{-1}

   Calculate the $\Delta S^\ominus$ of reaction form the figures given.

2. In a blast furnace, iron oxide is reduced by carbon monoxide and by carbon, depending on the temperature.

   a Write the equation that links ΔG to temperature.

   b The reduction of iron ore could be written thus:
   $$FeO(s) + CO(g) \rightarrow Fe(s) + CO_2(g)$$

   i Use the data to calculate the ΔG of reaction at 500 K;

	kJ mol^{-1}
$\Delta G_{formation}(FeO)$	= –231.8
$\Delta G_{formation}(CO)$	= –137
$\Delta G_{formation}(CO_2)$	= –395

ii Calculate the value for $\Delta G^\ominus$ at 1500 K, given that the reaction
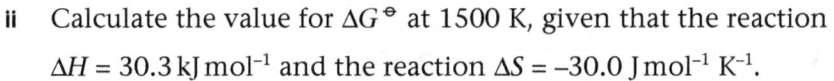
$\Delta H = 30.3\,\text{kJ}\,\text{mol}^{-1}$ and the reaction $\Delta S = -30.0\,\text{J}\,\text{mol}^{-1}\,\text{K}^{-1}$.

iii Is the reaction feasible at 500 K or at 1500 K? Explain your answer.

iv At what temperature would the reaction change from not being feasible to becoming just feasible? Show your calculations.

Unit 12 — ELECTRODE POTENTIALS AND FUEL CELLS

You need to know about **oxidation numbers** studies at AS level to understand this section.

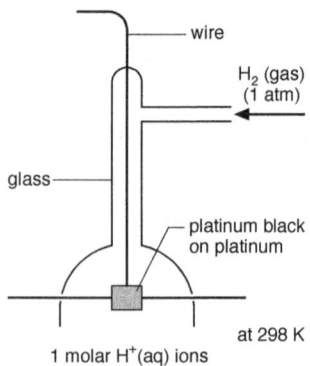

Fig. 12.1 The hydrogen electrode

Arranging $E^{\ominus}$ values in order produces the **electrochemical series**, which shows the relative oxidising or reducing powers of substances.

Powerful **oxidising agents** generally have $E^{\ominus}$ values greater than +1 V. **Example:** Fluorine, the strongest common oxidant.

$F_2(g) + 2e^- \rightleftharpoons 2F^-(aq)$

$E^{\ominus} = +2.87$ V

Powerful **reducing agents** generally have $E^{\ominus}$ values more negative than −1 V. **Example:** Potassium, the strongest common reductant.

$K^+(aq) + e^- \rightleftharpoons K(s)$

$E^{\ominus} = -2.92$ V

Oxidation is loss of electrons. **Reduction** is gain of electrons.

Oxidising agents (oxidants) oxidise other substances, so themselves are reduced. Oxidising agents gain electrons.

- Predicting whether a reaction will happen is a powerful idea in chemistry. Whether a reaction is **feasible** can be decided from **electrode potential** $E^{\ominus}$ values (if a **redox** reaction) or from **Gibbs energy** ΔG values for **any** reaction.

- When a piece of metal (called an **electrode**) is dipped into water, some metal atoms ionise. They leave **electrons** behind in the **metal** as they enter the water as **hydrated ions**. Example: Zinc: $Zn(s) \rightarrow Zn^{2+}(aq) + 2e^-$

  Some hydrated metal ions will find their way back to the electrode and **recombine** with the free electrons on it.

  $Zn^{2+}(aq) + 2e^- \rightarrow Zn(s)$

  In time the two reaction rates will equalise and **equilibrium** will be attained.

  $Zn(s) \rightleftharpoons Zn^{2+}(aq) + 2e^-$

  A **metal** electrode in contact with a solution of its **ions** is a **half cell**. There are electrons on the electrode and positive ions in solution, so the metal will have a negative charge and the solution a positive charge. There will be a **potential difference** (p.d.) between the solution and the metal electrode. This voltage is called the **electrode potential**.

- You cannot **measure** the electrode potential directly by connecting a voltmeter to a half cell. A voltmeter connecting **wire** that is dipped into the solution will have its **own electrode potential**. Therefore, the electrode potential of a metal must be compared with another electrode used as a **standard**. This arrangement is rather like measuring altitude from sea level rather than from the (unreachable) centre of the Earth. The standard **hydrogen electrode** is chosen as the standard and is assigned a potential of zero volts (see Fig. 12.1).

- The **standard electrode potential** of a half cell is the potential difference (measured in volts) between the half cell and a standard hydrogen electrode under standard conditions and when **no current flows**. Standard conditions are 298 K, 100 kPa, and solution concentration 1.00 mol dm^{-3}. The half equations for the zinc and the hydrogen electrodes are:

  $2H^+(aq) + 2e^- \rightleftharpoons H_2(g)$ 0.00 volts

  $Zn^{2+}(aq) + 2e^- \rightleftharpoons Zn(s)$ −0.76 volts compared with the H electrode.

  When a Zn half cell is connected to a hydrogen electrode and the circuit is completed, Zn produces electrons that flow to the hydrogen electrode. The potential has a **negative sign**.

  The **salt bridge** is a filter paper strip soaked in saturated aqueous KNO$_3$. As electrons flow through the external circuit, hydrated ions flow through the salt bridge to **complete** the circuit.

- You can **predict** whether a redox reaction is feasible by comparing electrode potentials.

  Example: Will Zn metal reduce Cu^{2+}(aq) ions? From a data book, the standard electrode potentials for the two redox reactions concerned are:

  $Cu^{2+}(aq) + 2e^- \rightarrow Cu(s)$ $E^{\ominus} = +0.34$ V

  $Zn^{2+}(aq) + 2e^- \rightarrow Zn(s)$ $E^{\ominus} = -0.76$ V

  The zinc half cell has the **more negative** potential, so electrons will flow **from the zinc** electrode **to the copper** electrode (which has a less negative i.e. more positive potential). The half equation for the zinc half cell moves to the **left** as it produces electrons; the half equation for the copper half cell moves to the **right** as it consumes electrons (see Fig. 12.2).

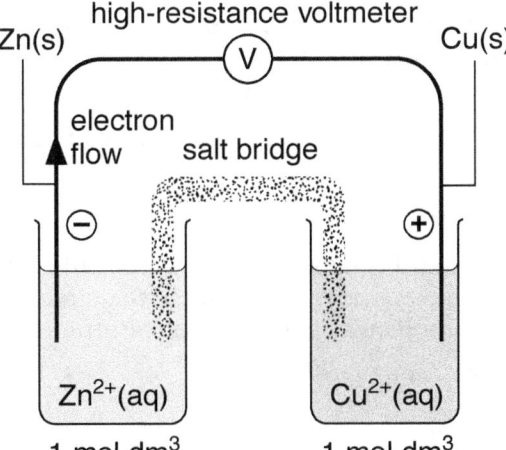

Fig. 12.2

$E^\ominus$reaction = $E^\ominus$right − $E^\ominus$left
= +0.34 − ⁻0.76
= +1.1 volts

> The reaction happens either when the reactants are **separated** into two half cells or when they are **mixed** together in a test tube.

The more negative $E^\ominus$ is the negative electrode

The more negative $E^\ominus$ produces e⁻
reactant

⊖ $Zn^{2+}(aq) + 2e^- \rightleftharpoons Zn(s)$ −0.76 V

⊕ $Cu^{2+}(aq) + 2e^- \rightleftharpoons Cu(s)$ +0.34 V
reactant

So this is feasible: $Zn(s) + Cu^{2+}(aq) \rightarrow Zn^{2+}(aq) + Cu(s)$
and the opposite is not feasible.

$Zn(s) \rightarrow Zn^{2+}(aq) + 2e^-$ (**oxidation**) $Cu^{2+}(aq) + 2e^- \rightarrow Cu(s)$ (**reduction**)

Adding the two half equations gives the full **redox** equation:

$Zn(s) + Cu^{2+}(aq) \rightarrow Zn^{2+}(aq) + Cu(s)$

The electrode potential of an **ion pair** is measured by placing a solution of concentration 1.0 mol dm⁻³ with respect to each ion in contact with a platinum electrode. Example: $Fe^{3+}(aq) + e^- \rightleftharpoons Fe^{2+}(aq)$ $E^\ominus = +0.77$ V

> You can apply **Le Chatelier's principle** to estimate the effect of non-standard conditions.

- Sometimes the electrode potentials predict that a reaction is **feasible**, but it does **not** occur. You should conclude that the reaction must have a high activation energy, so the rate is **low**. It is **kinetically stable**.

  Sometimes the standard electrode potentials predict that a reaction is not possible, but the reaction does occur in a test-tube. You must state that **non-standard conditions** are being used. Often the reactant concentrations are greater than 1.0 mol dm⁻³.

> **Disproportionation** may be explained by using $E^\ominus$ values.

- The **dry cells** you use in your personal mp3 player are electrochemical cells. A group of cells joined together is called a **battery**. These cells produce electricity by chemical redox reactions at their electrodes. **Oxidation** (loss of e⁻) happens at one electrode while **reduction** (gain of e⁻) happens at the other.

Unit 12 — ELECTRODE POTENTIALS AND FUEL CELLS

> An **electrochemical cell** consists of two connected half cells. The voltage of the cell is equal to the arithmetic difference between the $E^\ominus$ values.

- **Fuel cells** work by using oxygen (usually from the air) as the oxidising agent and a fuel (hydrogen or a hydrocarbon, e.g. methane) as the reducing agent. Most fuel cell electrodes consist of a metal foam that has a large surface area. Fuel cells are much more **energy efficient** than conventional power stations, and are portable and rugged.

- There are different types of fuel cells. Here is information about the more common alkaline electrolyte fuel cell. The hydrogen molecules lose electrons at one electrode:

  $H_2(g) + 2OH^-(aq) \rightarrow 2H_2O(l) + 2e^-$

  The electrons flow around the external circuit. Meanwhile the hydrogen ions flow between the electrodes to complete the circuit. At the oxygen/air electrode the oxygen molecules gain electrons in the presence of the water:

  $O_2(g) + 2H_2O(l) + 4e^- \rightarrow 4OH^-(aq)$

  The overall equation is:

  $2H_2(g) + O_2(g) \rightarrow 2H_2O(l)$

  Hydrogen is difficult to store as either it must be kept as high pressure liquid, adsorbed on the surface of a solid material, or absorbed within a solid material. Alternatively the hydrogen is stored in a hydrogen rich fuel such as methanol, natural gas, or petrol, which are converted into hydrogen gas by an onboard 'reformer'.
  The chemical energy released by the reaction of the fuel with the oxygen produces the voltage.

> **Storage cells** are recharged when an externally applied current reverses the chemical changes.

- In the car industry, scientists are developing fuel cell vehicles (FCVs), fuelled by hydrogen gas or by hydrogen-rich fuels. These vehicles will require roughly half the fuel per kilometre that similarly sized conventional vehicles require. Other technologies will increase efficiency in other ways.

  The hydrogen fuel cells only produce pure water from their exhaust, while hydrogen rich fuels only produce a small amount of carbon dioxide and other air pollutants.

  There are obstacles to the development of fuel cells for vehicles and with the storage and distribution of the fuel.

> The **specification** include these social and political points, but long discussion are unlikely.

- Though there are many advantages to fuel cells there are also disadvantages: (i) the hydrogen must be stored and transported safely, (ii) it may not be feasible to transport pressurised hydrogen liquid, (iii) the 'adsorber' or 'absorber' solids may not have a long enough lifetime; (iv) it costs to replace regularly the solids, (v) the disposal or recycling of the solids must be resolved; and the production costs may be too high, (vi) toxic chemicals may be required to produce the solids.

- Though there may be a political and social desire to move to a hydrogen economy there are many obstacles which include the ignorance that energy is needed to produce hydrogen and that fuel cells have a finite life.

  It may be that the 'hydrogen economy' may contribute to our energy needs but there must be public and political acceptance of hydrogen as a fuel, the technology required to handle and maintain the hydrogen systems must be resolved, and the initial manufacture of the hydrogen still requires energy.

- It is important to emphasis that in a hydrogen economy, the hydrogen must be produced and that requires energy.

TESTS

RECALL TEST

1. Define 'standard electrode potential'.

2. Draw and label the hydrogen electrode.

3. For these two electrode potentials, work out a balanced equation for a feasible reaction, and find which electrode is positive, and what the e.m.f. might be:

   $Ag^+(aq) + e^- \rightarrow Ag(s)$ +0.08 volts

   $Ni^{2+}(aq) + 2e^- \rightarrow Ni(s)$ –0.25 volts

4. Write the chemical equation for the reaction at:

   a the oxygen electrode in the hydrogen fuel cell,

   b the oxygen electrode in the hydrogen fuel cell

5. When the electrode potentials indicate a reaction is feasible but it does not occur, what must this be due to?

6. Why may a reaction occur in the test-tube even though the standard electrode potentials suggest it should not be feasible?

7. For rusting _____ and _____ are required.

8. Fuel cells use _____ or _____ and _____ as the fuel to produce electricity.

9. Give three advantages of fuel cells over conventional power stations.

10. A list of the metal electrode potentials in order is called the _____ _____

11. a What are the advantages of using fuel cells?

Unit 12 TESTS

11 b What are the technical obstacles to the adoption and large scale use of fuel cells in vehicles?

12 What are the non-technical obstacles to the adoption and large scale use of fuel cells in vehicles?

13 Give four ways that hydrogen may be stored in fuel cell vehicles.

14 Suggest how hydrogen may be produced on a large scale.

CONCEPT TEST

1 a Use the electrode potentials above to predict which halide ions may be

		$E^\ominus$
A	$Fe^{2+}(aq) + 2e^- \rightarrow Fe(s)$	−0.44
B	$Fe^{3+}(aq) + e^- \rightarrow Fe^{2+}(aq)$	+0.77
C	$Zn^{2+}(aq) + 2e^- \rightarrow Fe(s)$	−0.76
D	$Ni^{2+}(aq) + 2e^- \rightarrow Ni(s)$	−0.25
E	$I_2(aq) + 2e^- \rightarrow 2I^-(aq)$	+0.54
F	$Br_2(aq) + 2e^- \rightarrow 2Br^-(aq)$	+1.07
G	$Cl_2 + 2e^- \rightarrow 2Cl^-(aq)$	+1.36
H	$O_2(g) + 4H^+(aq) + 4e^- \rightarrow 2H_2O(l)$	+1.23

oxidised by oxygen to produce the halogens.

b In fact, if oxygen is mixed with the aqueous halide ions, no halogen is immediately produced. Suggest why this might be so.

2 Iron(II) ions are easily changed into iron(III) ions by many oxidising agents. Also, some reducing agents will reduce iron(III) ions to iron(II) ions.

a Will iron(III) ions oxidise iodide ions? Explain your answer.

b In practice, aqueous iron(II) ions will reduce iodine to produce iodide ions. Explain why this occurs.

c Which reagents, if any, in the equations above will oxidise iron(II) ions to iron(III) ions? Explain your answer.

d Use the electrode potentials above to explain which oxidation number is produced by the action of oxygen with water on iron.

e Here are the electrode potentials involved in rusting:

$$Fe^{2+}(aq) + 2e^- \rightleftharpoons Fe(s) \qquad E^\ominus = -0.44$$
$$\tfrac{1}{2}O_2(g) + H_2O(l) + 2e^- \rightleftharpoons 2OH^-(aq) \qquad +0.40$$

Write the equation for the reaction of the iron in water.

f Explain why zinc plating protects iron from rusting.

g Explain why nickel plating protects iron until the nickel is breached.

h Explain why iron(II) ions do not disproportionate to make iron(II) ions and iron metal.

3 The common alkaline dry cell is the battery used in bike lights and some mp3 players. At one electrode zinc metal reacts to make zinc(II)hydroxide under alkaline conditions, while at the other electrode, solid manganese(IV)oxide reacts in the presence of water to make solid manganese(III)hydroxide. A thick paste contains the required the water and alkali.

a Write an equation for the reaction of zinc in the presence of alkali. Include state symbols.

b Write an equation for the reaction of manganese(IV)oxide in the presence of water. Include state symbols.

c Which electrode would produce electrons when the cell is discharging?

d At which electrode is oxidation occurring?

e Write an overall equation for the discharging cell.

Unit 13 — TRANSITION ELEMENTS: THEORY

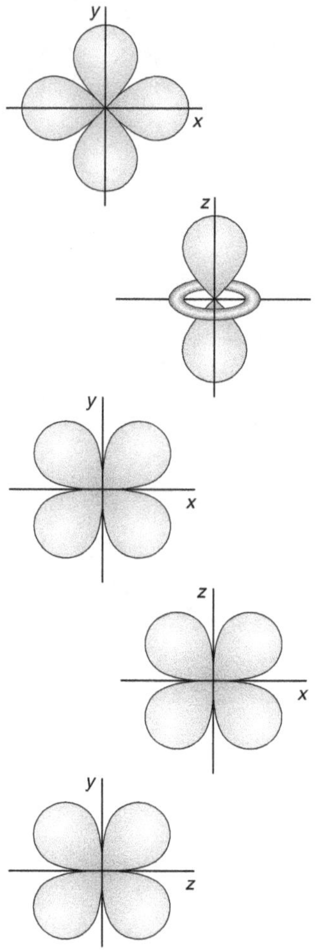

- The **d-block** elements occupy the **central part** of the periodic table in periods 4, 5, and 6. A set of **5 d orbitals** fill with a total of 10 electrons in the course of each period. You must concentrate on the metals of period 4 from **scandium** to **zinc**, in which the 3d orbitals fill from Sc = $3d^1 4s^2$ to Zn = $3d^{10} 4s^2$. Note that the 4s fills **before** the 3d.

d-block element	Symbol	Outer shell electronic configuration	Most important oxidation numbers
Scandium	Sc	$3d^1 4s^2$	+3
Titanium	Ti	$3d^2 4s^2$	+2 +3 +4
Vanadium	V	$3d^3 4s^2$	+2 +3 +4 +5
Chromium	Cr	**$3d^5 4s^1$**	+2 +3 +6
Manganese	Mn	$3d^5 4s^2$	+2 +4 +7
Iron	Fe	$3d^6 4s^2$	+2 +3
Cobalt	Co	$3d^7 4s^2$	+2 +3
Nickel	Ni	$3d^8 4s^2$	+2
Copper	Cu	**$3d^{10} 4s^1$**	+1 +2
Zinc	Zn	$3d^{10} 4s^2$	+2

Fig. 13.1

AS unit 1 chemistry will help you with this topic.

Four of the d orbitals are shaped like a four-leaf clover and one is shaped like a p orbital inside a doughnut ring (see Fig. 13.1).

- All the elements from Sc to Zn have an **inner shell** electronic configuration **$1s^2 2s^2 2p^6 3s^2 3p^6$**. These elements are listed in the table below:

  Note that the 4s has **higher energy** than the 3d. When ions form, the **4s** electrons are lost **before** the **3d** electrons. For example, Fe = $3d^6 4s^2$; Fe^{2+} = $3d^6 4s^0$ and Fe^{3+} = $3d^5 4s^0$. Also, the **d^5** and the **d^{10}** configurations have **extra stability**, hence Cr = $3d^5 4s^1$ rather than $3d^4 4s^2$ and Cu = $3d^{10} 4s^1$ rather than $3d^9 4s^2$. Similarly, Fe^{2+} ($3d^6 4s^0$) oxidises easily to Fe^{3+} ($3d^5 4s^0$) but it is difficult to oxidise Mn^{2+} ($3d^5 4s^0$) to Mn^{3+} ($3d^4 4s^0$) because it means breaking into the more stable $3d^5$ configuration.

  The table also shows the common oxidation numbers. Many elements **first** lose **all 4s** electrons to form an ion. For the elements up to Mn, the **maximum oxidation number** is equal to the **electron number** in the outer shell.

- A **transition element** is an element that forms at least one ion with a **partially filled d subshell**.
  They have **variable oxidation number** and the aquated ions are **coloured**. Sc has a single oxidation number only, Sc^{3+} = **$3d^0$** $4s^0$. The empty d subshell means that Sc is **not** a transition element. Zn has a single oxidation number only, Zn^{2+} = **$3d^{10}$** $4s^0$. The full d subshell means that Zn is **not** a transition element. Sc^{3+}(aq) and Zn^{2+}(aq) are **colourless**.

- **Colour** is determined by the element, the oxidation number, and the ligands (and sometimes the co-ordination number and shape).

- **Complexes** form when a central cation (or atom) forms **dative covalent** (co-ordinate) bonds by **accepting** electron pairs from ions (or molecules) called **ligands** (see Fig. 13.3). Ligands donate electrons from lone pairs or pi bonds.

Dative covalent bonds are covalents bonds where both electrons come from the same atom. Dative bonds may be called co-ordinate bonds.

Fig. 13.3 A complex

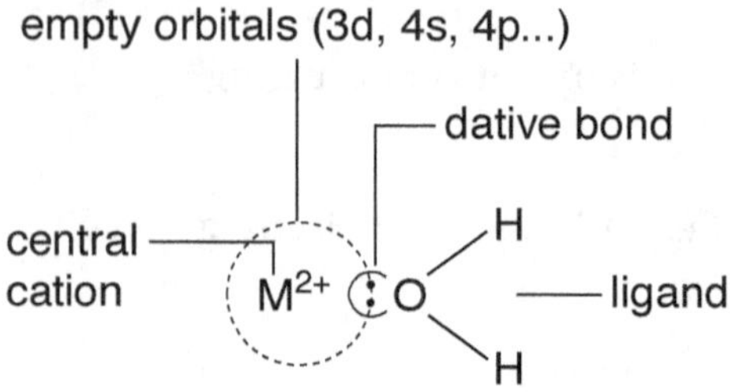

- Ligands that use **one** electron pair per molecule in complexes are called **monodentate** or unidentate ('one-toothed') ligands.

Formula	Name
NH_3	ammine
H_2O	aqua
OH^-	hydroxo
O^{2-}	oxo
Cl^-	chloro
CN^-	cyano
SCN^-	thiocyano
CO	carbonyl

Fig. 13.4 Monodentate ligands

- Polydentate (multidentate) ligands donate two or more pairs: **bidentate** ligands donate **2** pairs; **tetradentate 4** pairs; **hexadentate 6** pairs. The ligands you must know are listed in Figs 13.4, 13.5, 13.9.

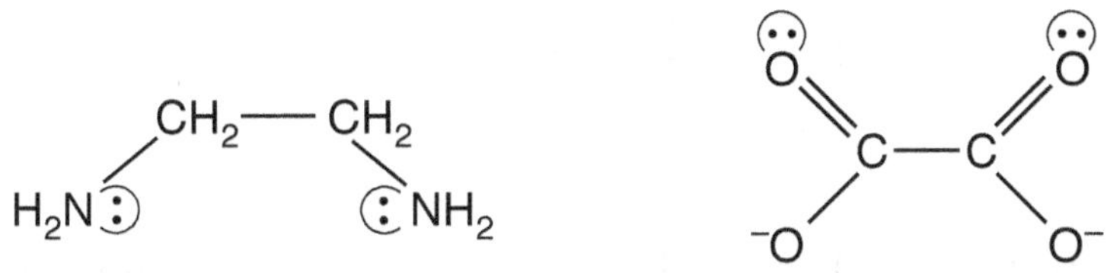

diaminoethane   ethane-1,2-dioate (oxalate) ion

Fig. 13.5 Bidentate ligands

- Complexes may have 6, 4, or 2 **electron pairs** binding with the **central ion** or **atom**. The number of electron pairs involved is called the **co-ordination number**. Co-ordination number 6 is the most common, 4 is less common, and 2 you will only meet with Ag^+ and Cu^+ ions.
- Complexes may be cationic, neutral, or anionic.

You must know the shape of complexes:

6 co-ordinate complexes are **octahedral**. **Example:** $[Cu(H_2O)_6]^{2+}$ is octahedral and neutral.

4 co-ordinate are usually **tetrahedral** with large ligands (occasionally **square planar** **Example:** $[CuCl_4]^{2-}$ is tetrahedral and anionic; $Ni(CO)_4$ is tetrahedral, and contains a neutral atom and 4 CO molecules; neutral $[Ni(NH_3)_2Cl_2]$ square planar so forms *cis-trans* stereoisomers (see below).

2 co-ordinate are always **linear.** **Example:** $[Ag(NH_3)_2]^+$ is linear and cationic.

- The **name** of a complex always starts with the ligands.
Examples:

$[Ag(NH_3)_2]^+$ diamminesilver(I);

$[CuCl_4]^{2-}$ tetrachlorocopper(II);

$[Cr(H_2O)_3(OH)_3]$ triaquatrihydroxochromate(III).

Unit 13 TRANSITION ELEMENTS: THEORY

- Complexes can show **isomerism**.

  Geometric isomerism is shown by Ni(NH$_3$)$_2$Cl$_2$; the Cl atoms may be next to (cis) or opposite each other (trans) (see Fig. 13.6).

Fig. 13.6

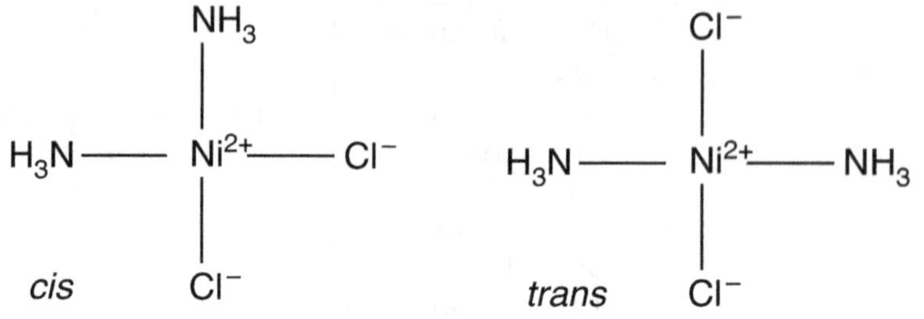

Fig. 13.7

Optical isomerism is shown by any complex accepting six lone pairs from 3 bidentate ligands e.g. [Ni(NH$_2$CH$_2$CH$_2$NH$_2$)$_3$]$^{2+}$ (see Fig. 13.7).

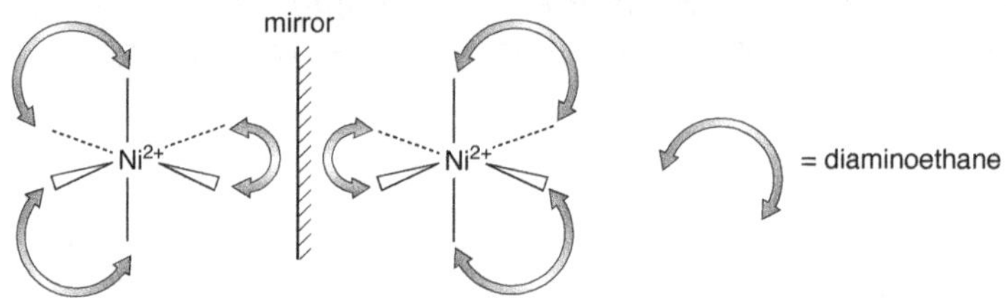

The Pt(II) complex *cis*-**platin** is used to treat cancer, while the *trans* form is not effective (see Fig. 13.8).

Fig. 13.8

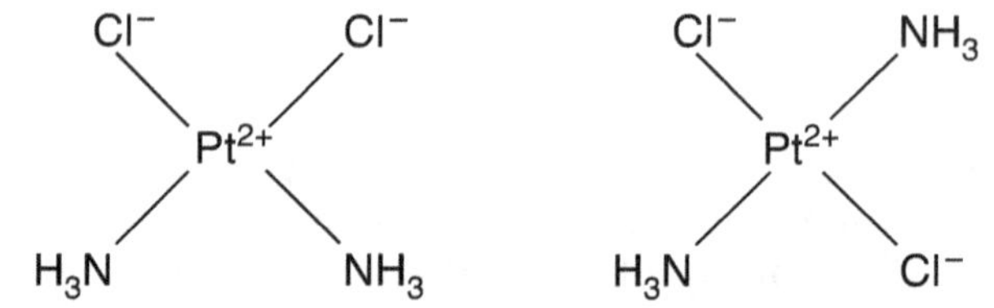

cis-platin is an anti-cancer drug   *trans*-platin is not

- The hexadentate ligand **EDTA** is used in shampoo, plant food, and some vitamins because it binds with free aqueous ions. It is also used to remove aquated ions from the bodies of people suffering from lead and cadmium poisoning. See Fig. 13.9

Fig. 13.9 EDTA

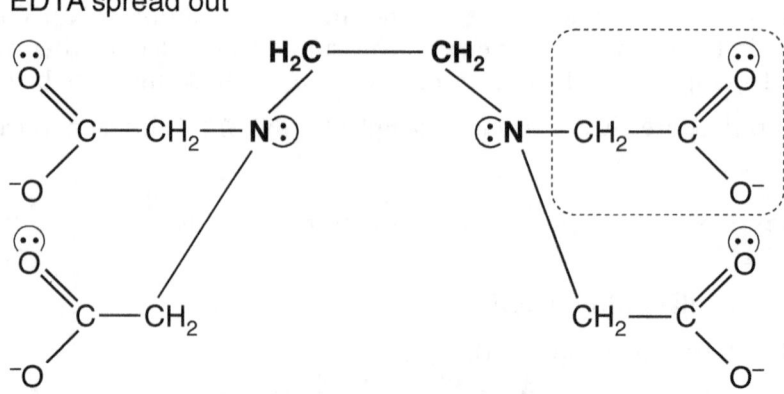

EDTA wraps around a metal ion

TESTS

RECALL TEST

1. What is meant by a 'd-block element'?

2. Which subshell is filled first and emptied first, the 3d or the 4s?

3. Write the full electronic configuration of:

   a Cu

   b Fe^{3+}

4. Why is it easy to oxidise Fe^{2+} to Fe^{3+} but difficult to oxidise Mn^{2+} to Mn^{3+}?

5. Suggest the shape of these complexes:

   a $[Fe(H_2O)_6]^{3+}$,

   b $[CoCl_4]^{2-}$ (this complex forms only one isomer)

   c $[PdCl_4]^{2-}$, (this complex forms two isomers).

6. What three factors determine the colour of a particular coloured complex?

7. Explain what is meant by a complex.

8. What is a ligand?

9. Explain what is meant by a 'monodentate ligand'.

77

Unit 13 TESTS

10 Give four examples of ligands.

11 Suggest which type of stereoisomerism is shown by:
   a $[Cr(H_2NCH_2CH_2NH_2)_3]^{3+}$

   b $[PdCl_2(NH_3)_2]^{2-}$.

12 What is the co-ordinate number of the complex $[Cu(H_2O)_4(H_2NCH_2CH_2NH_2)]^{2+}$?

CONCEPT TEST

1 Transition metals have characteristic properties.
   a State what is meant by 'transition elements'.

   b State which metals of the d-block (from Sc to Zn) are not transition metals.

   c State the three general properties of transition metals.

   d Which type of isomerism do transition metal complexes exhibit?

   e Which type of isomers exhibit cis-trans isomerism?

2 Complexes containing ligands are produced by transition metals and other metallic elements of the periodic table.
   a Explain what is meant by a 'ligand'.

b Complete the boxes to show the electronic configuration of iron and the iron ions (Ar) indicates an equivalent configuration.:

Fe atom (Ar) [ ]3d[ ] [ ] [ ] [ ]   [ ]4s

Fe^{2+} (Ar) [ ] [ ] [ ] [ ] [ ]   [ ]

Fe atom (Ar) [ ] [ ] [ ] [ ] [ ]   [ ]

c Suggest why is it easy to oxidise Fe^{2+} to Fe^{3+}, but difficult to oxidise Mn^{2+} ions?

3 Some ligands are bidentate.

  a Give an example of a bidentate ligand. Name it and give the structural formula.

  b Suggest how many of these bidentate ligands would fit around one chromium(III) ion.

  c State the type of isomerism exhibited by the complex in part **b**.

  d Suggest the shape of hexaaquachromium(III) ions.

Unit 14 — TRANSITION ELEMENTS: REACTIONS

- You must know the **colours** of the aqueous cations of Cu^{2+}, Co^{2+}, Fe^{2+}, and Fe^{3+} (see Fig. 14.1).

Aqueous cations	Colour
Cr^{3+}(aq)	purple
Mn^{2+}(aq)	colourless
Fe^{2+}(aq)	blue-green
Fe^{3+}(aq)	brown
Co^{2+}(aq)	pink
Ni^{2+}(aq)	green
Cu^{2+}(aq)	blue
Zn^{2+}(aq)	colourless

Fig. 14.1

- You will find transition metal chemistry easier to understand if you learn to recognise the **types of reaction**: acid–base, redox, ligand substitution, precipitation, and thermal decomposition. The examiner will test your understanding as well as your recall of these ideas.

- **Acid–base** reactions are when a complex gains or loses a **proton** (H^+ ion). Losing a proton is called **deprotonation**. Example:

  $[Fe(H_2O)_6]^{3+}(aq) + 3OH^-(aq) \rightarrow [Fe(OH)_3(H_2O)_3](s) + 3H_2O(l)$

  Alternatively, you are allowed to write this more simply as:

  $Fe^{3+}(aq) + 3OH^-(aq) \rightarrow [Fe(OH)_3](s)$

> **Oxidation** is loss of electrons. **Reduction** is gain of electrons.

- **Redox** reactions are when an element changes oxidation state. **Example:**

  $Cr_2O_7^{2-}(aq) + 3Zn(s) + 14H^+(aq) \rightarrow 2Cr^{3+}(aq) + 3Zn^{2+}(aq) + 7H_2O(l)$

  The oxidation number of Cr reduces from +6 to +3. The oxidation number of Zn increases from 0 to +2.

  The examiner is likely to link electrode potentials to transition metal questions.

> **Oxidising agents** (oxidants) oxidise other substances, so themselves are reduced. Oxidising agents gain electrons.

- **Ligand substitution** (exchange or displacement) is when one ligand replaces another. **Example:** Aqueous Cu^{2+} ions are pale blue. Adding aqueous NH_3 forms a deep blue solution.

  $[Cu(H_2O)_6]^{2+}(aq) + 4NH_3(aq) \rightarrow [Cu(NH_3)_4(H_2O)_2]^{2+}(aq) + 4H_2O(l)$

  NB The reaction

  $[Cu(H_2O)_6]^{2+}(aq) + 2OH^-(aq) \rightarrow [Cu(H_2O)_4(OH)_2](s) + 2H_2O(l)$

  appears to be ligand replacement, but in fact it is an acid–base reaction in which $OH^-(aq)$ is protonated by water ligands in the complex.

  Thermal decomposition occurs as with group 2 compounds.

- Forming insoluble hydroxide **precipitates** helps to identify many transition metals. The precipitating agent is either (i) the strong base **NaOH(aq)** which contains a high concentration of $OH^-(aq)$ ions, or (ii) **aqueous NH_3** which contains small concentrations of $OH^-(aq)$ and high concentrations of $NH_3(aq)$.

  $NaOH(s) + H_2O(l) \rightarrow Na^+(aq) + OH^-(aq)$

  $NH_3(aq) + H_2O(l) \rightleftharpoons NH_4^+(aq) + OH^-(aq)$

  Generally, adding a **few drops** of either NaOH(aq) or NH_3(aq) to an aqueous transition metal cation forms a **hydroxide precipitate**.

  $M^{2+}(aq) + 2OH^-(aq) \rightarrow M(OH)_2(s)$

$M^{3+}(aq) + 3OH^-(aq) \rightarrow M(OH)_3(s)$

- As with most of the corresponding s- and p-block metal compounds, transition metal **nitrates**, **sulphates**, and **chlorides** are **soluble** in water. The insoluble silver(I) and copper(I) halides are the only important exceptions. **Carbonates** and **hydroxides** are **insoluble**.

- You must know the colours of the aqueous ions and of the precipitates of Cu^{2+}, Co^{2+}, Fe^{2+}, and Fe^{3+}.

Cation	+ drops NaOH(aq)	+ drops NH_3(aq)	In excess NaOH(aq)	In excess NH_3(aq)
Cr^{3+}	green ppt	green ppt	green solution	does not dissolve
Mn^{2+}	grey* ppt	grey* ppt	does not dissolve	does not dissolve
Fe^{2+}	green* ppt	green* ppt	does not dissolve	does not dissolve
Fe^{3+}	brown ppt	brown ppt	does not dissolve	does not dissolve
Ni^{2+}	green ppt	green ppt	does not dissolve	violet solution of $[Ni(NH_3)_6]^{2+}$
Cu^{2+}	blue ppt	blue ppt	does not dissolve	deep blue solution of $[Cu(NH_3)_4(H_2O)_2]^{2+}$
Zn^{2+}	white ppt	white ppt	colourless solution of $[Zn(OH)_4]^{2-}$	colourless solution of $[Zn(NH_3)_4]^{2+}$

ppt = precipitate

The copper(II) ions dissolve in excess NH_3(aq) to form an **ammine complexes**.

When a hydroxide **dissolves** in excess **NH_3(aq)**, an ammonia complex is formed by **ligand substitution**. Example:

$[Cu(OH)_2(H_2O)_4](s) + 4NH_3(aq) \rightarrow [Cu(NH_3)_4(H_2O)_2]^{2+}(aq) + 4H_2O(l)$
    pale blue   dark blue

- Ligand substitution occurs when concentrated HCl(aq) (or NaCl) are added to aqueous complexes.

$[Cu(H_2O)_6]^{2+}(aq) + 4Cl^-(aq) \rightarrow 6H_2O(l) + [CuCl_4]^{2-}(aq)$

Note the co-ordination number **change** from 6 to 4. Only four of the large Cl^- ions can fit around the small Cu^{2+} ion. You should **know** this example of $[CuCl_4]^{2-}$(aq) (yellow).

Aqueous **thiocyanate ions** will form a blood-red solution with Fe^{3+}(aq) by ligand substitution (it really does look like real blood).

$[Fe(H_2O)_6]^{3+}(aq) + SCN^-(aq) \rightarrow [Fe(SCN)(H_2O)_5]^{2+}(aq) + H_2O(l)$

Ammonia NH_3(aq) forms $[Ag(NH_3)_2]^+$(aq) which is the active ingredient in Tollen's reagent.

> **Cyanide** ion CN^-(aq) forms $[Ag(CN)_2]^-$(aq) used in silver plating.

Iron ions are in a complex, called Haem, within **haemoglobin**. Oxygen molecules form a loose complex with haem allowing oxygen to be transported around the body. Carbon monoxide ligands will substitute the oxygen molecules and form a permanent dative covalent bond with the haem which decreases oxygen transport. This is why carbon monoxide poisoning, often due to poorly maintained gas appliances, kills people every year.

Unit 14 — TRANSITION ELEMENTS: REACTIONS

- The following reactions, colours and formulae are specified by your exam board. Other reactions may be stimulus material in exams when the examiner wants to test your understanding.

  To help learn the colours of the ions, you could use coloured pencils to shade colour near each formula.

- Blue green **aqueous iron(II) ions** form a green precipitate with drops of dilute sodium hydroxide or aqueous ammonia:

  $Fe^{2+}(aq) + 2OH^-(aq) \rightarrow Fe(OH)_2(s)$

  This green precipitates turn brown due to **oxidation** by the **air**. You do not have to recall this, but the examiner may mention it:

  $4Fe(OH)_2(s) + O_2(g) + 2H_2O(l) \rightarrow 4Fe(OH)_3(s)$

- Yellow brown **aqueous iron(III) ions** form a brown precipitate with drops of dilute sodium hydroxide or aqueous ammonia:

  $Fe^{3+}(aq) + 3OH^-(aq) \rightarrow Fe(OH)_3(s)$

- Blue **aqueous copper(II) ions** form a blue precipitate with drops of dilute sodium hydroxide or aqueous ammonia:

  $Cu^{2+}(aq) + 2OH^-(aq) \rightarrow Cu(OH)_2(s)$

  Blue aqueous copper(II) ions with excess dilute ammonia form a deep blue solution:

  $[Cu(H_2O)_6](aq) + 4NH_3(aq) \rightarrow [Cu(NH_3)_4(H_2O)_2]^{2+}(aq) + 2H_2O(l)$

  Blue aqueous copper(II) ions with excess concentrated chloride ions (in concentrated hydrochloric acid or sodium chloride solution) produces a yellow solution:

  $[Cu(H_2O)_6]^{2+}(aq) + 4Cl^-(aq) \rightarrow CuCl_4^{2-}(aq) + 6H_2O(l)$

- Pink **aqueous cobalt(II) ions** form a blue precipitate with drops of dilute sodium hydroxide or aqueous ammonia:

  $Co^{2+}(aq) + 2OH^-(aq) \rightarrow Co(OH)_2(s)$

  Pink aqueous cobalt(II) ions with excess concentrated chloride ions (in concentrated hydrochloric acid or sodium chloride solution) produces a blue solution:

  $[Co(H_2O)_6]^{2+}(aq) + 4Cl^-(aq) \rightarrow CoCl_4^{2-}(aq) + 6H_2O(l)$

- Equilibrium constants may be applied to complex formation. The stability constant, K_{stab}, is the equilibrium constant for the formation of a complex ion in a solvent from its constituent ions. Consider:

  $Cu^{2+}(aq) + 4Cl^-(aq) \rightleftharpoons CuCl_4^-(aq)$

  K_{stab} is similar to K_c:

  $K_{stab} = [CuCl_4^-(aq)] / [Cu^{2+}(aq)] [Cl^-(aq)]^4$

  You maybe required to deduce K_{stab} from a given equation.
  You may also be asked to deduce the units of K_{stab}.

  In the this example,
  the units of K_{stab} = (mol dm^{-3}) / (mol dm^{-3}) x (mol dm^{-3})^4
  = mol^{-4} dm^{12}, or dm^{12} mol^{-4}.
  The larger the K_{stab} the more stable the complex ion.

TESTS
RECALL TEST

1. State the colours of these aqueous ions:

   Fe^{2+} :

   acidified Fe^{3+}:

   Co^{2+}:

   Cu^{2+}:

2. What types of reaction are these?

   a $5MnO_4(aq) + 8H^+(aq) + 5Fe^{2+}(aq) \rightarrow Mn^{2+}(aq) + 5Fe^{3+}(aq) + 4H_2O(l)$

   b $[Zn(H_2O)]^{2+}(aq) + 2OH^-(aq) \rightarrow [Zn(OH)_2(H_2O)_4](s) + 2H_2O(l)$

   c $[Zn(OH)_2(H_2O)_4](s) + 2OH^-(aq) \rightarrow [Zn(OH)_4]^{2-}(aq) + 2H_2O(l)$

   d $[Cu(H_2O)_6]^{2+}(aq) + 4Cl^-(aq) \rightarrow 6H_2O(l) + [CuCl_4]^{2-}(aq)$

3. Indicate which of these aqueous ions react as described:

	Fe^{2+}	Fe^{3+}	Co^{2+}	Cu^{2+}
Forms ppt with NaOH(aq)				
Dissolves in excess NaOH(aq)				
Forms ppt with NH_3(aq)				
Dissolves in excess NH_3(aq)				
ppt oxidises in air				

4. Write balanced equations for these reactions:

   a the addition of aqueous ammonia to iron(II) sulfate solution.

   b aqueous copper(II) ions with in excess NaCl(aq)

   c aqueous cobalt (II) ions with excess HCl(aq)

   d aqueous copper(II) sulfate with excess aqueous ammonia.

5. Explain what is meant by ligand substitution.

6. Give an expression and units for K_{stab} for this reaction:

   $[Ni(H_2O)_6]^{2+}(aq) + 4Cl^-(aq) \rightarrow NiCl_4^{2-}(aq) + 6H_2O(l)$

Unit 14 TESTS

CONCEPT TEST

i an aqueous solution of 3.16 g dm^{-3} potassium manganate(VII), KMnO$_4$, labelled solution P,

ii an aqueous solution of ethanedioic (oxalic) acid, H$_2$C$_2$O$_4$, of unknown concentration, labelled Q,

ii an aqueous solution of 7.865 g dm^{-3} vanadium(III) chloride, VCl$_3$, labelled R,

iii aqueous bench sulfuric acid.

1 Iron is a typical transition metal.

   a Iron(II) sulphate dissolves in water to make a pale green solution. On addition of NaOH(aq) a green precipitate forms, which changes colour when left in the air.

      i Give the formula of the first precipitate.

      ii Why would the green precipitate change colour in air?

      iii Iron(II) ions react with zinc to make iron and zinc(II) ions. Write a balanced ionic equation for this redox reaction.

   b Iron(II) ions will react with manganate(VII) ions (permanganate) to produce manganese(II) ions.

      i Write a half equation for the oxidation of iron(II) ions.

      ii Write a half equation for the reduction reaction.

      iii Hence write an ionic equation for the oxidation of iron(II) ions by manganate(VII) ions.

2 Pale blue aqueous copper(II) ions, [Cu(H$_2$O)$_6$]$^{2+}$, with drops of dilute aqueous NH$_3$, will form a blue precipitate, Cu(OH)$_2$, which redissolves in excess NH$_3$(aq) to form a dark blue solution, [Cu(NH$_3$)$_4$(H$_2$O)$_2$]$^{2+}$.

   a State the type of reaction that is illustrated by the formation of a blue precipitate.

   b State the type of reaction for the formation of a dark blue solution.

   c Suggest how could the dark blue solution be converted back to form a pale blue solution?

   d How could the aqueous copper(II) ions be converted into yellow tetrachlorocopper(II) ions (tetrachlorocuprate(II) ions)?

3 Use these electrode potentials to help you answer the questions.

		$E^\ominus$
A	$Fe^{2+}(aq) + 2e^- \rightarrow Fe(s)$	–0.44
B	$Fe^{3+}(aq) + e^- \rightarrow Fe^{2+}(aq)$	+0.77
C	$Zn^{2+}(aq) + 2e^- \rightarrow Zn(s)$	–0.76
D	$MnO_4^-(aq) + 8H^+(aq) + 2e^- \rightarrow Mn^{2+}(aq) + 4H_2O(l)$	+1.52
E	$VO_2^+(aq) + 4H^+(aq) + e^- \rightarrow VO^{2+}(aq) + 2H_2O(l)$	+1.00
F	$VO^{2+}(aq) + 2H^+(aq) + e^- \rightarrow V^{3+}(aq) + H_2O(l)$	+0.34
G	$V^{3+}(aq) + e^- \rightarrow V^{2+}(aq)$	-0.26
H	$V^{2+}(aq) + e^- \rightarrow V(s)$	-1.20

Which reagents will:

a reduce VO_2^+ to V^{2+}?

b oxidise V^{2+} to VO_2^+ or VO_3^- ions?

c oxidise Fe^{2+} to Fe^{3+} ions?

d change Fe^{3+} to Fe^{2+} ions?

4 Here are the stability constant for the formation for the two copper(II) complexes indicated:

$[Cu(CN)_4]^{3-}$ $K_{stab} = 2.0 \times 10^{27}$ (see below for units).

$[Cu(NH_3)_2]^+$ $K_{stab} = 1.0 \times 10^{11}$ (see below for units)

which is from:

$Cu^+(aq) + 2NH_3(aq) \rightleftharpoons [Cu(NH_3)_2]^+$.

a Write an equation for the formation of the $[Cu(CN)_4]^{3-}$ complex:

b What units would the K_{stab} for $[Cu(NH_3)_2]^+$ have?

c Which complex is the more stable?

d If copper(I) sulfate was added to a mixture of equimolar aqueous potassium cyanide and aqueous ammonia which complex would form?

Unit 15 — SYNOPTIC: LINKING CHEMISTRY TOGETHER

- If you are taking **A2** level papers, then you will have to face the general, or **synoptic papers,** that draw together the whole specifications under the heading **Unifying Concepts**. These titles may sound frightening and it may seem that the examiner can change topics within a question without warning. However, if you fully understand the basics of chemistry and stay calm, you will find these questions an interesting challenge. The examiners are only allowed to ask questions on topics that are on the specifications.

  You will face questions that may seem to have nothing to do with the specifications you studied. If you cannot immediately think of the relevant chemistry, **don't panic**. Pause and let your mind wander the specifications. Does the question remind you of an idea you have studied?

- Your examiners have agreed the following aims for **general or synoptic questions**. They state:

  "... Candidates should be able to:
  (i) bring together knowledge, principles and concepts from different areas of chemistry, including experiment and investigation, and apply them in a particular context, expressing ideas clearly and logically and using appropriate specialist vocabulary;
  (ii) use chemical skills in contexts which bring together different areas of the subject. ..."

- So be ready to:

  Use facts, ideas, and practical knowledge, from **any** part of the specifications.
  Apply these in **new** situations.
  Suggest new ideas based on known ideas.
  Communicate your answer clearly, using **chemical** terms.

> It is crucial that you understand the **command words** that the examiner uses. Look at the introduction for explanations of these words.

- The examiner can jump from one topic to another in any part of the specifications. Here are some **common connections**:

  Rates connected to **mechanisms** e.g. S_N2.
  Thermochemistry with any organic or inorganic reactions.
  Rates, equilibrium, and **thermochemistry** together applied to an **industrial process** that you have not met before.
  Bond enthalpies applied to **organic chemistry** (perhaps the stages of a mechanism) and so to **reaction rates**.
  Electrode potentials with inorganic reactions, particularly those involving **transition metals** or group 7 elements.
  An **organic mechanism** applied to a new situation.
  Mechanism applied to an **inorganic** setting e.g. a covalent chloride.
  Organic chemicals used as **ligands** with d-block metal ions.
  Mole calculations mixed in with any of the above.

- A **worked example**: This question deals with various aspects of copper chemistry.

  (a) Copper ions, in a complex, are used in a test for a functional organic group.
      (i) Name the test. (1 mark)

A	$Cu^+(aq) + e^- \rightleftharpoons Cu(s)$	$E^\ominus = +0.52$ V	D	$I_2(aq) + 2e^- \rightleftharpoons 2I^-(aq)$	$E^\ominus = +0.54$ V
B	$Cu^{2+}(aq) + e^- \rightleftharpoons Cu^+(aq)$	$E^\ominus = +0.15$ V	E	$Cu^{2+}(aq) + I^-(aq) + e^- \rightleftharpoons CuI(s)$	$E^\ominus = +0.87$ V
C	$Cu^{2+}(aq) + 2e^- \rightleftharpoons Cu(s)$	$E^\ominus = +0.34$ V	F	$S_2O_8^{2-}(aq) + 2e^- \rightleftharpoons 2SO_4^{2-}(aq)$	$E^\ominus = +2.01$ V

  (ii) Name the functional group that gives a positive result in the test.

(iii) Give the physical state and formula of the copper-containing species produced.

(b) Here are some standard electrode potentials ($E^\ominus$) in volts (V).
   (i) Study A and B above and use the electrode potentials to explain the disproportionation of $Cu^+(aq)$ and to write a balanced equation for the reaction.

   (ii) Study B and F and write a balanced equation for the reaction that happens when the half cells are connected together. State the cell e.m.f. and identify the positive electrode.

   (iii) When aqueous copper(II) ions are mixed with aqueous iodide ions, this reaction occurs:
   $2Cu^{2+}(aq) + 4I^-(aq) \rightarrow 2CuI(s) + I_2(aq)$
   With reference to the electrode potentials, suggest why this reaction occurs.

   (iv) Explain why there is no observable change when aqueous persulphate ions $S_2O_8^{2-}(aq)$ are mixed with aqueous potassium iodide.

(c) Why is it effectively impossible to change copper(II) compounds into copper(III) compounds?

(total 18 marks)

● The answers discussed:
(a) (i) Here you need to think of a distinctive organic test that uses copper ions. Recalling all organic tests, you should think of the Fehling's test which uses a copper ammine complex.

   (ii) The name of the functional group is aldehyde.

   (iii) The formula of the copper compound made is Cu_2O, copper(I) oxide, which is a red solid.

(b) (i) A and B are electrode potentials. You need to recall that electrons flow from the more negative half cell to the more positive i.e. from B to A. Equilibrium B moves to the left and A to the right. Reversing B and adding to A produces:

   B $Cu^+(aq) \rightarrow Cu^{2+}(aq) + e^-$ (reversed)
   A $Cu^+(aq) + e^- \rightarrow Cu(s)$
   ───────────────────────────────
       $2Cu^+(aq) \rightarrow Cu^{2+}(aq) + Cu(s)$ (note the electrons cancel)

   (ii) F has the more positive electrode potential and so will have the positive electrode. Electrons will flow from equilibrium B to F. Equilibrium B moves to the left and F to the right. B produces only one electron while F consumes 2. **Doubling** B, reversing it, and adding to F gives

   B $2Cu^+(aq) \rightarrow 2Cu^{2+}(aq) + 2e^-$ (doubled and reversed)
   F $S_2O_8^{2-}(aq) + 2e^- \rightarrow 2SO_4^{2-}(aq)$
   ───────────────────────────────
       $2Cu^+(aq) + S_2O_8^{2-}(aq) \rightarrow 2Cu^{2+}(aq) + 2SO_4^{2-}(aq)$ (electrons cancel)

Unit 15
SYNOPTIC: LINKING CHEMISTRY TOGETHER

(iii) The examiner used the word **suggest** to check whether you really knew that the electrode potentials can explain why the reaction happens. It is always worth seeing whether an immediate and obvious explanation works first. Then, if it does not work, you must think further.

One answer would be that the $E^{\ominus}$ value for D is more negative than the value for E. Equilibrium D will produce electrons (I⁻ is oxidised to I), and E will accept electrons (Cu^{2+} is reduced to Cu^+). Therefore, the reaction as written is feasible.

(iv) When you check the $E^{\ominus}$ values for the relevant equilibria (D and F), you conclude that the reaction is feasible. The question states that it does not occur, so you must look for another explanation. In this case, suggest that the activation energy may be high so the rate is so low that the reaction does not appear to happen.

(c) The examiner has changed the subject (note the letter change from part (b) to (c), hinting at a new topic). Many other metals have a maximum oxidation state of +2, so why do they not oxidise further? Think about what would be involved in achieving the change:

$Cu^{2+}(aq) \rightarrow Cu^{3+}(aq) + e^-$

The answer is that the third ionisation energy is very high (due to breaking into a closed electron shell) and requires the input of an enormous amount of energy unlikely to be recovered through hydration of the 3+ ion. The overall reaction would therefore be extremely endothermic.

- A **structured** approach to **problem solving** is to **first** extract the data and **then** look for combinations of data that allow you to use an appropriate equation.

- When a question requires more than a simple recall answer, especially answers requiring a **longer answer** then follow these points. With practice you will speed up.

  Read the question.

  Consider all the technical **words** and their meaning.

  Connect the **data** together.

  Connect the ideas to the different parts of the specifications.

  If you do not immediately see the connection then just **keep going**. The topics must be in your specification.

  Read the question again.

  Glance at the number of **marks**.

  Make **deductions**.

  Put your answer into **order**.

  Read the question again.

  Start to **write** your answer.

- With long answers just work through the question in order as the wording usually contains many points.

- Writing longer answers requires some people to practice. If your answers are confused then make yourself practice rewriting answers to past exam questions until

TESTS

SYNOPTIC EXAM-STYLE QUESTIONS

1 Chlorine is found in many useful chemicals. Often the amount of chloride in a solution is determined by titration with aqueous silver nitrate using potassium dichromate solution as an indicator.

   a Write an ionic equation for the reaction between chloride ions and silver nitrate solution.

   (1)

   b 161.4 g of an unknown alkali earth chloride is dissolved in water, then made up to 250 cm^3 with water. 25 cm^3 of the solution is titrated with 0.01 mol dm^{-3} AgNO$_3$(aq), using potassium chromate solution as an indicator; 22.95 cm^3 is needed to change the indicator colour. By working out its RAM, identify the unknown element.

   (4)

   c Other than by using aqueous silver ions, how could the presence of aqueous iodide ions be detected?

   (2)

   d Suggest why the theoretical lattice energy of iron(III) chloride is very different from the lattice energy determined by experimentation.

   (2)

   e When concentrated hydrochloric acid is added to a solution of copper(II) chloride the solution changes colour. Give the formula of the copper-containing species produced. Explain simply why the different complexes produce different colours.

   (3)

   f Chlorine compounds react in different ways. Explain simply *how* each of these chlorine compounds react with the given reagent:

   i sodium chloride with acidified potassium manganate(VII) (potassium permanganate)

   ii chloroethane with aqueous potassium hydroxide

Unit 15 TESTS

    iii chloroethane with aqueous potassium hydroxide

    iv methylbenzene with chlorine and ultraviolet light

    v methylbenzene with chlorine and aluminium chloride

(5)

g Why will aqueous sodium hydroxide react with $C_6H_5CH_2Cl$, but not with $ClC_6H_4CH_3$?

(3)

(Total 20 marks)

2 Hydrogen cyanide production is a major use of methane. The hydrogen cyanide may be used to extract gold, or manufacture methyl 2-nitrilepropenenoate. In addition HCN is a weak acid.

a Hydrogen cyanide production takes place by this reaction:

$$2CH_4(g) + 2NH_3(g) + 3O_2(g) \rightarrow 2HCN(g) + 6H_2O(g)$$

The actual industrial conditions are 1000 °C with a Pt/Rh catalyst. Use the average bond enthalpies left to suggest a standard enthalpy change for this reaction.

Average bond enthalpy (kJ mol^{-1})	
H-C	412
H-N	388
H-O	436
O=O	496
C≡N	890

(3)

b State, with a short reason, the effect of increasing the temperature on:

    i the position of equilibrium

    ii the forward reaction rate

    iii the reverse reaction rate

(3)

c In terms of rate and yield, explain why such a high temperature is used.

(2)

d Why is high pressure not used?

(2)

e The first stage of gold extraction, which uses aerated sodium cyanide, is by this reaction:

$$4Au(s) + 8CN^-(aq) + O_2(g) + 2H_2O(l) \rightarrow 4[Au(CN)_2]^-(aq) + 4OH^-(aq)$$

   i State the oxidation state of the gold in the product.

   ii Name the complex $[Au(CN)_2]^-$.

   iii What type of bonding is in the gold complex $[Au(CN)_2]^-$?

   iv Once the gold solution is concentrated, enough zinc metal is used to liberate the free metal. Suggest an ionic equation for this reaction.

(4)

f Explain why the standard enthalpy change of neutralisation of HCl with KOH is 57.2 kJ mol^{-1}, but that of HCN with KOH is 11.7 kJ mol^{-1}.

(2)

g The structural formula of the monomer of a transparent polymer is shown in Fig. 15.1.

   i Draw the repeating unit of the polymer.

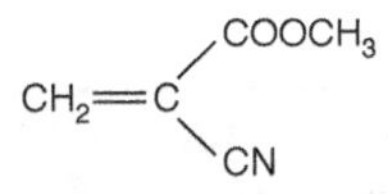

$$CH_2=C\begin{smallmatrix}COOCH_3\\CN\end{smallmatrix}$$

Fig. 15.1

(2)

   ii The polymer is made under similar conditions to poly(phenylethene) and poly(ethene). Suggest the conditions that are used to make the transparent polymer.

(2)

(Total 20 marks)

Unit 16

EXPERIMENTAL SKILLS

- There is not enough space here to give detailed descriptions of the methods used in practical chemistry. The intention of these notes is to show you how to use or describe **practical techniques** in written exams or coursework practicals.

- In OCR A level chemistry, 20% of the marks are for the **coursework** (that is 10% for the AS and 10% for the A2 coursework). Three types of tasks will be set: **Qualitative, Quantitative, and Evaluative tasks.**

  The tasks are carried out in class under strict supervision. The exam papers must be kept secure. During the exam you must not talk or communicate with anyone (except those supervising you) in any way. Treat the session like a written paper exam. A fresh set of coursework tasks are set each year. You may have up to three tries at each task, so you may attempt up to nine coursework tasks for AS and nine for A2. Only the best mark in each skill area counts.

  Qualitative tasks are out of 10 marks. They test your powers of observations, and simple interpretation. Often the tasks are likely to be based on inorganic chemistry (group 2 and 7 chemistry for AS, and transition metals for A2) or perhaps organic reactions (for both AS and A2).

  Quantitative tasks are out of 15 marks. They will be based on calculation based practicals. One task is likely to be based on a titration (for AS it is likely to be an acid-base titration, and for A2 a redox titration). Enthalpy or rates practicals are other possibilities. You may have to calculate yield or percentage error.

  Evaluative tasks are out of 15 marks. It is likely that no actual practical work will be set, so expect just a written paper. Evaluation is a difficult as there are so many aspects to practical work. You may have to evaluate a moles based method, or a rates or enthalpy practical. Again you may have to calculate yield or percentage error.

- When you are to make **observations** then note down **everything** you see and sense. Note the number of marks and always write more points than the marks. You may have to explain inorganic or organic reactions, so ensure you can recall and understand the reactions.

- **Percentage error** = the accuracy of a piece of apparatus divided by the measurement multiplied by 100.

  Percentage error = accuracy/measurement x 100

  Example: a mass of 2.05 grams is measured using a balance that is accurate to 0.01 grams. The % error = 0.01 /2.05 x 100 = 0.049%.

- **Percentage yield** = actual mass produced divided by the theoretical mass times 100.

  Percentage yield = actual/theoretical x 100

  Example: Aspirin ($CH_3COOC_6H_4COOH$) is made from salicylic acid (HOC_6H_4COOH). If 10.00 grams of salicylic acid produces 9.45 grams of Aspirin, what is the percentage yield?

  $(CH_3CO)_2O(l) + HOC_6H_4COOH(s) \rightarrow CH_3COOC_6H_4COOH(s) + CH_3COOH(l)$

  Moles of salicylic acid = mass / Mr = 10.00 / 138 = 0.07246376812 moles,
  Ratio is 1:1,
  So theoretical mass = moles x Mr = 0.07246376812 / 180 = 13.043478261 g.
  Percentage yield = 9.45/13.043478261 x 100 = 72.45% = 72.5% (significant figures limited by the 9.45 gram mass).

> When calculating molar amounts calculate which substance is in excess, and determine which substance is limiting the yield.

> Usually an organic compound undergoing reaction is not in excess; in redox reactions, the acid is in excess.

- **Precision** is important. Many instruments can usually provide highly accurate data, but if you are asked to state a measurement to only 1 °C or to 2 significant figures **then do so**. The **least accurate** measurements (those with the least number of significant figures) **limit** the overall experimental accuracy. When measuring using a graduated piece of apparatus (e.g. a measuring cylinder) then measure to half a division unless told otherwise. If data is correct to **2** significant figures, you **cannot** give the result of a calculation to **3** significant figures. However while carrying out calculations store all the figures, do not round down until the end of the calculation, otherwise you may get a rounding error.

- **Accuracy and reliability** are important. You must know the difference.
  Accuracy is how close a piece of data is to the true value. Accurate data would also have to be precise.
  Reliability is whether the data may be trusted. Reliable data will be repeatable. Data may be measured to four significant figures, but if a method produces data with a wide range then it is unreliable.

  Example: These readings are precise, accurate and reliable: 25.231, 25.230, 25.232.

  Example: These readings are precise, accurate but not reliable 20.225, 24.279, 25.232, assuming the average is around the true value.

  Example: These readings are reliable but not accurate or precise: 25, 25, 25, assuming the true value is 28.

- Accuracy improves if the experiment is repeated and an **average** calculated.

- Always add **units** to your results and when showing your answers in calculations. Always show some steps in your calculations as they earn marks.

- Consider these practical points when during a **qualitative** or **quantitative task**, but also during an **evaluative task**:
  To ensure all reactant molecules have **full contact** with each other, mixtures of liquids must be **mixed** by stirring or shaking.
  Stir liquids as soon as they are poured together.

  The term **heat strongly** is associated with the thermal decomposition of compounds. You should **heat gently** when trying to speed up a reaction without boiling away solvent water or causing decomposition. Use an **oil bath** to heat above 100 °C. A bare flame causes localised hot-spots so a gauze helps. An **electric heater** is especially useful when heating flammable liquids.

  When asked to measure a **volume**, and when given a choice of glassware, choose one which is as large as the volume required, or larger. Beakers cannot be used to measure volumes precisely. (E.g. 45 cm³ volume is measured in 50 cm³ cylinder).

  The **melting point** of a compound (up to about 250 °C with an oil bath) is determined by using a capillary tube and thermometer (see Fig. 16.1).

  The **boiling point** is measured by placing a thermometer **just above** the surface of the boiling liquid. A flask is a convenient container (see Fig. 16.2).

  Heating or boiling **under reflux** speeds up a reaction without losing volatile (readily vaporised) reactants or products. A flask with a vertical condenser attached is used (see Fig. 15.3).

- **Rate** experiments usually involve measuring time. Remember to measure the time carefully. Be careful to measure the correct **start time** and the **stop time**. Rates practicals are very sensitive so measure everything carefully. Any of the **five main factors** (concentration, surface area, temperature, pressure and catalyst) may influence the rate, but focus on the first three.

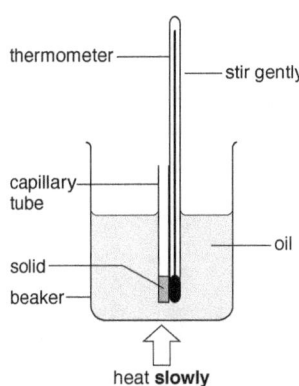

Fig. 16.1 Melting point apparatus

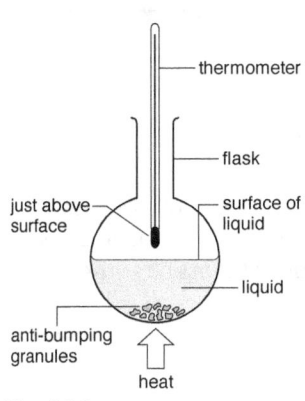

Fig. 16.2

Unit 16 — EXPERIMENTAL SKILLS

> Stir liquids as soon as they are poured together.

- The **enthalpy change** for a reaction taking place in solution can be calculated from **temperature changes** and measurements of **mass/volume** (grams/cm^3 of water) and **amount** (moles of reactants). You must remember to note the temperatures you read **before the start** of the reaction and **at the end**. Remember also that **inaccuracies** are introduced by **heat loss** to the container or to the air (by conduction or evaporation). **Incomplete reaction** may be caused by insufficient stirring.

- Details of **safety points** are usually included in assessed practicals and they are often examined in written papers. The examiners assume that safety glasses and lab coats are worn, but **mention them anyway**. Always give a reason for your safety point.

  Watch out for **flammable** substances and include the phrase use **no naked flames** in your practical instructions.

  Where poisonous solids are used add the obvious phrase '**do not eat**' and include the use of protective **gloves**.

  Volatile and gaseous **poisons** must be used in a **fume cupboard**.

- When suggesting improvement, perhaps in an **evaluative task**, then link a criticism to an improvement.

> Remember also that solutions must be made up accurately.

  If there may be a temperature loss, then suggest that the temperature be kept constant by putting a beaker in a **water bath** with a **lid**.

  Ensure that the solutions are completely mixed.

  To decrease the percentage error, **double** the volumes or masses used. This also helps rates practicals to be more accurate.

  To improve accuracy suggest that the practical be **repeated**, an **average** taken, that **anomalous** (outliers) are left out, or that the practical be repeated until **consistent** results are obtained.

  Consider the factors that may influence the **rate**. Suggest how to improve the accuracy of the measurement of volume and temperature. Discuss how to make the surface area of a catalyst the same for all experimental runs. Suggest how to make the time longer as this would have a lower percentage error.

  Look at the practical points on the previous pages.

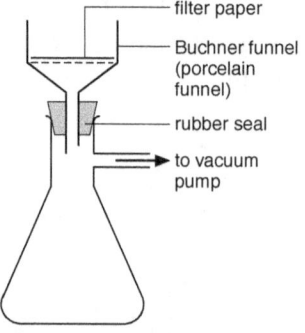

Fig. 16.3 In recrystalisation a Buchner funnel may be used to fast filter the hot solvent

- Use **recrystallization** to purify a compound which is more soluble in hot solvent than cold. Examiners may ask you to state the stages in recrystallization or may ask for explanations. The stages are:
  (i) **Dissolve** the impure substance (which is effectively a mixture) in the **minimum** amount of **hot solvent** (inorganic compounds in water; organic compounds in water, or ethanol, or a liquid hydrocarbon);
  (ii) **Filter** the hot mixture under reduced pressure, to remove insoluble impurities (see Fig. 16.3);
  (iii) **Cool** the filtrate slowly to form crystals (fast cooling traps impurities within the crystals);
  (iv) **Filter** the cold mixture, to remove soluble impurities. The **residue** is the pure substance.

TESTS

TESTS

RECALL TEST

1. What is meant by:

    a Accuracy,

    b Precision,

    c Reliability.

2. What must you remember to do when carrying out:

    a a Qualitative task,

    b a Quantitative task,

    c a Evaluative task,

3. How may the following practicals be made more accurate?

    a rates practical,

    b an enthalpy practical,

    c a titration practical,

    c a gas volume practical,

Unit 16

TESTS

CONCEPT TEST

1. During a titration of ethanoic acid against potassium hydroxide, the titre was found to be 23.45 cm³ ethanoic acid against a 10.00 cm³ KOH aliquot. If the alkali was 0.60 mol dm⁻³, what is the concentration of the ethanoic acid?

$$CH_3COOH(aq) + KOH(aq) \rightarrow CH_3COO^-K^+(aq) + H_2O(l)$$

2. Often iodine is titrated against thiosulfate. If a titre of 26.85 cm³ potassium thiosulfate solution (which contains $2S_2O_3^{2-}$ ions), 0.020 mol dm⁻³, reacted with 50.0 cm³ iodine, dissolved in potassium iodide, what would be the iodine concentration?

$$2S_2O_3^{2-}(aq) + I_2(aq) \rightarrow S_4O_6^{2-}(aq) + 2I^-(aq)$$

3. Silver chloride is less soluble than silver chromate, so the chloride ion concentration may be determined by titrating aqueous silver nitrate into a chloride ion solution. A few drops of yellow potassium chromate, here being used as an indicator, will produce a red silver chromate precipitate (mixed in with the yellow). If the aliquot was 25.0 cm³ and the 0.010 mol dm⁻³ aqueous silver nitrate titre was 23.45 cm³, what was the chloride ion concentration?

4. Ammonium sulphate, $(NH_4)_2SO_4$, is used as a fertiliser. A 24.0 g sample of the fertiliser was added to excess NaOH(aq) and the resulting ammonia gas absorbed in 1 dm³ sulphuric acid, 0.10 mol dm⁻³. 10.0 cm³ of the resultant solution was titrated with 0.010 mol dm⁻³ NaOH. 29.10 cm³ was required. Calculate the purity of the ammonium sulphate.

5. When 25.00 cm³ of a copper ion solution was mixed with excess potassium iodide solution, the iodine produced required 40.10 cm³ of 0.0200 mol dm⁻³ sodium thiosulfate solution. The reactions were these:

$$2Cu^{2+}(aq) + 4I^-(aq) \rightarrow 2CuI(s) + I_2(\text{in aqueous KI})$$

$$2S_2O_3^{2-}(aq) + I_2(aq) \rightarrow S_4O_6^{2-}(aq) + 2I^-(aq)$$

Calculate the concentration of the copper ion solution.

6 Benzene (M_r = 78) may be nitrated with concentrated nitric acid at 40 °C to make nitrobenzene (M_r = 123):

$$C_6H_6(l) + HNO_3(l) \rightarrow C_6H_5NO_2 + H_2O(l)$$

a A student used 10.0 grams benzene and obtained 9.0 grams nitrobenzene. What percentage yield was produced?

b Explain why the same student using the same amount could appear to make a yield of 120% later the same day (assuming the mathematics was correct).

7 Ethyl ethanoate was be prepared using ethanol and ethanoic acid. 10.0 ml of the acid is mixed with 10.00 grams of ethanol, and 15.00 grams of the ester is made. The density of ethanoic acid is 1.049 g cm^{-3}.

$$CH_3COOH(aq) + CH_3CH_2OH(aq) = CH_3COOCH_2CH_3(aq) + H_2O(l)$$

a Calculate the percentage yield.

b Calculate the percentage error for:

i the volume reading where a 10 cm^3 measuring cylinder (accurate to 0.1 cm^3) was used to measure the ethanoic acid.

ii the mass reading where a balance was used (accurate to 0.01 grams) which was used to measure the ethanol.

c To how many significant figures may the answer to **a** be given? Give your reasoning.

d Suggest three ways that the method may be inaccurate and how the method could be improved.

Inaccuracy 1:

Improvement 1:

Inaccuracy 2:

Improvement 2:

Inaccuracy 3:

Improvement 3:

ANSWERS

UNIT 1

RECALL TEST
1. The p orbitals on each carbon atom overlap with the adjacent p orbitals to merge into one delocalised ring of pi electrons.
2. The enthalpy of delocalisation may be determined by comparing the benzene hydrogenation with the hydrogenation of a simple alkene (e.g. cyclohexene). The benzene hydrogenation will be found to be less than three times the cyclohexene hydrogenation. The difference is due to the enthalpy of delocalisation. An energy-level diagram would illustrate the ideas effectively.
3. Same as Fig. 1.5.
4. Same as Fig. 1.6.
5. a $C_6H_6 + HNO_3 \rightarrow C_6H_5NO_2 + H_2O$.
   b $C_6H_6 + Cl_2 \rightarrow C_6H_5Cl + HCl$.
   c $C_6H_6 + Br_2 \rightarrow C_6H_5Br + HBr$.
   d $C_6H_6 + CH_3COCl \rightarrow C_6H_5COCH_3 + HCl$.
6. In hydrated aluminium chloride the aluminium atoms would be surrounded by water molecules, so lone pairs from the Cl would not be able to join with the Al. The Al may no longer act as a halogen carrier.
7. Benzene with ethene and HCl.
8. Explosives and dyes.
9. sodium phenoxide, C_6H_5ONa
10. a Concentrated nitric acid and concentrated sulfuric acid.
    b Tin and concentrated hydrochloric acid and heat.
    c Hydrochloric acid (or sulphuric acid) and sodium nitrite, at 5 °C.
    d Phenol (or 2-naphthol) with aqueous NaOH.
    e Aqueous NaOH.
    f Aqueous dilute nitric acid.

CONCEPT TEST
1. a Same as Fig. 1.6.
   b Electrophilic substitution.
   c The electron density of the π bonds of the C=C in alkenes are localised so cyclohexene easily reacts with bromine, whereas the delocalised electron density of the π bonds in benzene are delocalised so benzene is resistant to bromination.
   d Phenol much more reactive than benzene due to electron-pair donation to the delocalised ring from the oxygen p-orbital.
2. a B is $CH_3C_6H_4NO_2$.
   b Tin and concentrated hydrochloric acid.
   c $CH_3C_6H_4N_2^+$.
   d An azo dye.
   e You should have drawn an azo dye. The N=N link in the middle contains two bonds. Similar to Fig. 19.10, but with a methyl group on one benzene ring and the OH⁻ group on the other.
   f With iron(III) chloride solution phenol will produce a violet colour.
3. a Chloromethane and aluminium chloride, heat.
   b Chlorine and UV light.
   c Aqueous sodium hydroxide.
   d Concentrated nitric acid and concentrated sulfuric acid.
   e Aqueous sodium hydroxide, room temperature.

UNIT 2

RECALL TEST
1. Dipole–dipole forces (and Van der Waals forces).
2. nucleophilic addition, alcohol, aldehydes, carboxylic acid.
3. Similar to Fig. 2.2, but propanone replaces the ethanal molecule.
4. Propanal. Propanal molecules only have dipole–dipole forces and Van der Waals forces, whereas propan-1-ol molecules also have hydrogen bonding.
5. ethanoic acid.
6. a Reagent is 2,4-dinitrophenylhydrazine; produces a brightly coloured solid.
   b The solid is purified by recrystalisation. The melting point of the solid is determined. The melting point indicates the carbonyl compound (by referring to a data book that has the melting points of the derivatives).
7. a Propan-1-ol is heated and distilled with acidified potassium dichromate solution.
   b Propan-2-ol is heated (under reflux) with acidified potassium dichromate solution.
8. a Reduction, propan-1-ol.
   b Reduction, propan-2-ol.
9. a $CH_3CH_2CHO + 2[H] \rightarrow CH_3CH_2CH_2OH$.
   b $CH_3COCH_3 + 2[H] \rightarrow CH_3CHOHCH_3$.
   c $CH_3CHO + (O) \rightarrow CH_3COOH$.
   d $CH_3CHO + (O) \rightarrow CH_3COOH$.
10. Tollen's reagents.
11. Propanone is warmed with $NaBH_4$(aq) (or $LiAlH_4$) then acidified.
12. a Acidified potassium dichromate, $Cr_2O_7^{2-}/H^+$, and heated (under reflux if large quantities).
    b Acidified potassium dichromate, $Cr_2O_7^{2-}/H^+$, and heated and distilled.
    c Acidified potassium dichromate, $Cr_2O_7^{2-}/H^+$, and heated (under reflux if large quantities).

CONCEPT TEST
1. a Y must be a ketone as it reacts with 2,4-DNP but not with Tollen's reagent.
   b Y must be propanone as it is a three carbon ketone.

c X must be an alcohol. X must be propan-2-ol as only that alcohol may be oxidised to form a ketone.
d By using a strong reducing agent: $NaBH_4(aq)$ and acidified (or dry $LiAlH_4$ and then acidity).

2 a The unknown compound must be warmed gently with Tollen's reagent (ammoniacal silver nitrate solution). A silver mirror (of black precipitate) will form with the aldehyde but not the ketone.
b By recrystalisation.
c The melting point of the purified solid indicates the identity of the carbonyl compound.

3 a Use a pure sample in a mass spectrometer. On the mass spectrum the large peak to the right indicates the molecular mass of the molecule.
b Both the infra red spectra for aldehydes and ketones have a peak of absorbance between 1640 -1750.
  b i Both propanal and propanone will show a chemical shift due to the -CH_3 group between 0.8 and 2.0.
  ii Propanone, CH_3COCH_3, has only one peak, because there is only one proton environment namely due to the -CH_3 group.
  iii The C=O group does not produce a chemical shift with a ^1H NMR, because there is no proton in that group, though the proton in the -CHO group does between 9 and 10.
  iv The -CH_2- peak is split because it is next to the -CH_3 group. It will split into 4 sub-peaks due to the n+1 rule. (There are there H in -CH_3 and 1+1 = 4).
  ii $CH_3CHO + (O) \rightarrow CH_3COOH$.

UNIT 3

RECALL TEST

1 a $CH_3COOH + NaOH \rightarrow CH_3COO^-Na^+ + H_2O$.
b $CH_3COOH + NaHCO_3 \rightarrow CH_3COO^-Na^+ + H_2O + CO_2$.
c $2CH_3COOH + 2Na \rightarrow 2CH_3COO^-Na^+ + H_2$
d $CH_3COOH + CH_3CH_2OH \rightarrow CH_3COOCH_2CH_3 + H_2O$.
e $CH_3COOH + 4[H] \rightarrow CH_3CH_2OH + H_2O$.

2 Blue damp litmus turns pink/red, $NaHCO_3$ gives off CO_2 bubbles, Na(s) gives off H_2 bubbles, ethanol with concentrated H_2SO_4 produces a smell (often fruity), PCl_5 fizzes/produces steamy fumes (of HCl).

3 Propanoic acid, ethanol, and conc. H_2SO_4.

4 $CH_3COOCH_2CH_3 + H_2O \rightarrow CH_3COOH + CH_3CH_2OH$.

5 Salicylic acid with ethanoic acid anhydride. Warm gently. Neutralise in dilute sodium hydrogen carbonate. (recrystalize to purify)

6 a $(CH_3CO)_2O + H_2O \rightarrow 2CH_3COOH$.
b $(CH_3CO)_2O + CH_3CH_2OH \rightarrow CH_3COOCH_2CH_3 + CH_3COOH$.
c $CH_3CH_2Cl + 2NH_3 \rightarrow CH_3CH_2NH_2 + NH_4Cl$.
d $CH_3COCl + CH_3CH_2NH_2 \rightarrow CH_3CONHCH_2CH_3 + HCl$.
e $CH_3NH_2 + HCl \rightarrow CH_3NH_3^+ + Cl^-$.
f $NH_3 + CH_3Br \rightarrow HBr + CH_3NH_2$.
g $CH_3NH_2 + CH_3Br \rightarrow HBr + (CH_3)_2NH$.

7 a Dry ethanol at room temperature/ warm gently.
b Aminomethane at room temperature.
c Aqueous NaOH.
d Aqueous NaOH or $NaCO_3$ or Na(s).

8 $H_2SO_4(aq)$ or NaOH(aq), and heat under reflux.

CONCEPT TEST

1 a P: CH_3COOH;
   Q: $(CH_3CO)_2O$;
   R: $CH_3COOCH_2CH_3$;
   S: $CH_3COO^-Na^+$;
   T: CH_3CH_2OH; U: CH_3NH_2;
b i $CH_3COOH + NaOH \rightarrow CH_3COO^-Na^+ + H_2O$ (or use Na or $NaHCO_3$).
  ii $(CH_3CO)_2O + CH_3CH_2OH \rightarrow CH_3COOCH_2CH_3 + CH_3COOH$.
  iii $CH_3COOCH_2CH_3 + NaOH \rightarrow CH_3COO^-Na^+ + CH_3CH_2OH$.
c i Room temperature/warm gently.
  ii Room temperature with NaOH(aq), Na(s) or $NaHCO_3(s)$
  iii NaOH(aq) Heat under reflux.

2 a $CH_3NHCH_2CH_2NHCH_3$.
b i Add bromine water. With an unsaturated fat or oil it will go from brown to colourless (due to the C=C).
  ii trans fatty acids are associated with heat disease.
  iii $HOCHCH(CH_2)_7COOH$
c i *Either* add CH_3CH_2OH with conc. H_2SO_4, warm, neutralise with $NaHCO_3$ and smell a fruity smell of an ester if the original substance was CH_3COOH, *or* add $NaHCO_3(s)$ and see effervescence.
  ii *Either* add CH_3CH_2OH with conc. H_2SO_4, warm, neutralise with $NaHCO_3$ and smell a fruity smell of an ester if the original substance was CH_3COOH, *or* add $NaHCO_3(s)$ and see effervescence.

3 a ethanoic acid anhydride, $(CH_3CO)_2O$, with phenyl amine, $C_6H_5NH_2$, *or* phenyl ammonium chloride, $C_6H_5NH_3Cl$,
b $(CH_3CO)_2O + C_6H_5NH_2 \rightarrow CH_3CONHC_6H_5 + CH_3COOH$
c By recrystalisation

UNIT 4

RECALL TEST

1. $HOOC(C_6H_4)COOH$ and ethane-1,2-diol, $HOCH_2CH_2OH$.
2. a $[-OC(C_6H_4)COOCH_2CH_2O-]_n$.
   b $[-HN(CH_2)_6NHOC(CH_2)_8CO-]_n$.
   c $[-OCH(CH_3)CH_2CO-]_n$.
   d $[-HNC(R)HCO-]_n$.
3. Any triol, e.g. $CH(CH_2OH)_3$.
4. a $H_2N(CH_2)_6NH_2$ and $HOOC(CH_2)_4COOH$.
   b $H_2N(C_6H_4)NH_2$ and $HOOC(C_6H_4)COOH$.
   c $H_2NC(R)HCOOH$.
5. a Terylene is used to make fibres.
   b Nylon 6,6 is used as a fibre in clothes, in ropes to tie up ships, and hard-wearing parts of home appliances.
   c Kevlar is used in bullet-proof vests and to protect tyres and canoes.
6. hydrolyse.
7. a $^+H_3NCH(R)COO^-$.
   b $^+H_3NCH(R)COOH$.
   c $H_2NCH(R)COO^-$.
8. a Addition.
   b Polypeptide.
   c Polyester.
   d Polyamide.
   e Polyester.
9. Optical isomerism occurs when two isomers have the same structure, but have a different arrangement in space by being mirror images of each other that cannot be superimposed.
10. * indicates the chiral carbon atoms: $HOCH_2C^*HClCH_2C^*H(CH_3)COOH$.
11. Optical isomers rotate the plane of polarised monochromatic light in opposite directions. To measure this a polarimeter may be used.
12. You must draw clear 3D diagrams. The bonds should suggest the 109° bond angle around the central C or you lose a mark. Around the C on the -COOH the bond angles should be 120 degrees. Similar to Fig. 4.5.

CONCEPT TEST

1. a 1: $H_2N(CH_2)_6NH_2$ and $HOOC(CH_2)_4COOH$.
   2: $H_2NC(CH_3)HCOOH$, $H_2NC(CH_2CH_3)HCOOH$.
   3: $HOOC(C_6H_4)COOH$ and $HOCH_2CH_2OH$.
   4: CH_2CH_2.
   b 1: polyamide/nylon; 2: polypeptide; 3: polyester; 4: polyalkane.
   c 1: hydrogen bonding; 2: hydrogen bonding; 3: dipole–dipole; 4: Van der Waals forces.
2. a i $[-OC(C_6H_4)CONH(C_6H_4)NH-]_n$.
   ii $[-HN(CH_2)_5CO-]$
   iii $[-OC(C_6H_4)COOCH_2CH_2O-]_n$.
   b Condensation/addition elimination.

UNIT 5

RECALL TEST

1. a A pharmaceutical is drug or medicine that has a beneficial effect on the body chemistry.
   b A chiral centre is a carbon atom with four different groups on it. It shows optical activity.
   c a tertiary is a molecule that contains a C atom which his attached to a halogen atom and three carbon atoms.
2. To minimise side effects, to decrease cost, and to reduce risk from litigation.
3. Increased cost.
4. Enzymes; bacteria; chiral catalysts; chemical chiral synthesis; natural chiral molecules.
5. (Natural) L-amino acids or sugars.
6. Recrystallization.
7. Thin layer chromatography.
8. Cyclic strained molecules, and the use of reagents fixed to a polymer support with reactants flowing over them. Supercritical CO_2.

CONCEPT TEST

1. a Optical isomerism. The chiral carbon is indicated by the C*:

   $HOCH_2C(OH)(CH_3)C^*H(OH)CH_3$

   The isomers should be shown as mirror images is shown in Fig.

   b
   i $HOOCC(OH)(CH_3)COCH_3$
   ii $OHCC(OH)(CH_3)COCH_3$ or written backwards: $CH_3COC(OH)(CH_3)CHO$
   iii $HOCH=C(CH_3)CH=CH_2$, $HOCH_2(C=CH_2)CH=CH_2$.
   iv $NaBH_4(aq)$ has no effect.

   iv A molecule with three ester links:

   $CH_3COOCH_2C(OOCCH_3)(CH_3)CH(OOCCH_3)CH_3$

2.
   a 1,2,3-tribromopropane, or 1,2,3-trichloropropane.
   b A molecule with three ester links:

   $CH_3COOCH(COCOCH_3)CHOCOCH_3$

3. a Chlorine and UV light.
   b $H_2NCHCOOH$. As the reagent is alkaline, the correct answer is $H_2NCHCOO^- NH_4^+$.
   c Methanol, CH_3OH, with concentrated sulfuric acid.
   d R is an amino acid.
   e None. It does not have a C=C so does not show E-Z isomerism. It also does not have a carbon with four different groups on it so it does not show optical isomerism.
   f $^+H_3NCHCOO^-$
   g $-[-NHCHCO-]-$

UNIT 6

RECALL TEST

1 a Bromine water turns colourless.
b Orange $K_2Cr_2O_7$ goes green. Addition of conc. ethanoic acid (and $H_2SO_4(l)$) produces a sweet smell. PCl_5 vigorously produces white fumes. (Purple $KMnO_4$ goes colourless.)
c Blue litmus turns pink/red. Any carbonate (e.g. $NaHCO_3(s)$) produces effervescence. Addition of ethanol (and $H_2SO_4(l)$) produces a sweet smell. PCl_5 vigorously produces white (HCl) fumes.
d 2,4-dinitrophenylhydrazine produces a brightly coloured solid.
e Tollen's produces a silver mirror/black precipitate. (Fehling's produces a red solid.)
f Add pink litmus. It turns blue with an amine.
g Add aqueous NaOH, warm gently, neutralise with dilute nitric acid. Add $AgNO_3(aq)$ and shake. A cream precipitate forms.
h Phenol will turn blue litmus pink/red. Bromine water will decolourise and produce oily droplets that smell of antiseptic. ($FeCl_3$ forms a violet solution.)
2 a Blue flame.
b Yellow, very sooty flame.
3 bond vibrations.

CONCEPT TEST

1 a i P and R; brown bromine water is decolourised, detecting an alkene group >C=C<.
ii P; produces a silver mirror/black precipitate.
iii P only; turns from orange to green, detects the aldehyde group -CHO.
b P. On the proton NMR spectrum there are multiple peaks between 7.5 and 8 due to a benzene ring. There is absorbance between 9 and 10 indicating a -CHO group. (the small peaks are due to the near by aromatic ring). There is absorbance between 4.5 and 6 indicating a -C=CH- group. There are two peaks due to the other -C=CH- proton splitting the peak into two (n+1=1+1=2). Only P has these groups.
2 a 122 (largest mass/charge ratio).
b -COOH, carboxylic acid, because there is absorbance in the range 2500–3300 cm^{-1}. It is not an alcohol OH group as the chart lacks absorbance at 3200–3550 cm^{-1}.
c X must be benzoic acid C_6H_5OOH.
d The C-13 NMR chemical shifts would be: between 190 and 220 due to the C in the -CHO group; between 5 and 55 due to the -CH_3 group and there will be a cluster of peaks between 110 and 165 due to the aromatic ring

UNIT 7

RECALL TEST

1 The rate equation links rate to the concentrations of the chemicals which control the reaction rate, e.g. rate = $k[HI]^2$.
2 Order is the sum of the powers in the rate equation.
3 $mol^{-1}\,dm^3\,s^{-1}$.
4 s^{-1}.
5 $mol\,dm^{-3}\,s^{-1}$.
6 Order suggests the number of particles involved in the rate determining step.
7 Half life is the time taken for the reactant concentration to halve.
8 Colour, colorimeter; pH, pH meter; electrical conductivity, conductivity meter; polarised light, polarimeter; gas volume, gas syringe.
9 Rate [a] [b]
  1 1 1
  2 2 1
  3 3 1
  4 1 2
  9 1 3
10 See Fig. 12.1.
11 The reaction is zero order. The gradient tells you that rate is constant.
12 The reaction is first order. The gradient tells you the value for k.
13 The reaction is second order. The gradient tells you the value for k.
14 $E_a = 2830\,J\,mol^{-1}$.
15 a Concentration against time and rate against concentration.
b Rate against concentration and half life against time.
c Rate against concentration squared.

CONCEPT TEST

1 a Propanone = 1st order; I_2 = zero order; H^+ ions = 1st order.
b I_2 is a reactant, but it is not in the rate determining step.
c Rate = $k[CH_3COCH_3]^1\,[H^+]^1$.
d Two.
e The acid must be a catalyst as the acid is in the rate equation but is not used up by the reaction. (In fact, as H^+ ions are made by the reaction it is called an autocatalyst).
f The rate determining step in a mechanism is the slowest step that controls the overall reaction rate.
g CH_3COCH_3 and H^+ only. (Only they appear in the rate equation.)
2 a You cannot write the rate equation as order is only determined experimentally.
b *Either* by using a colorimeter which follows the $[MnO_4^-(aq)]$ *or* by using a pH meter which follows the $[H^+]$.

c Half life is the time taken for the reactant concentration to halve.
  d X is zero order.
  e A graph of rate against [Y] should be a straight line if [Y] is first order. The graph will start at the origin.

UNIT 8

RECALL TEST

1 'Dynamic equilibrium' is when in a reversible reaction the concentrations of reactants and products do not change, but the reactants are continually producing products and the products produce reactants.
2 Le Chatelier's principle states that if the conditions of a system at equilibrium are changed then the equilibrium position will shift to resist the change.
3 a Kc is unchanged. Right.
   b Kc is unchanged. Left.
   c Kc is unchanged. Right.
   d Kc is not changed as the enthalpy is zero. No change.
   e Kc is unchanged. No change in the position of equilibrium (as catalysts do not change the position of equilibrium).
4 a Kc is unchanged. Left.
   b Kc decreases. Left.
   c Kc is unchanged. Right.
5 a Kc decreases. Left.
   b Kc is unchanged. No change in the position of equilibrium as there are the same number of gas molecules on each side of the equation.
6 $mol\,dm^{-3}$.

7 a $\dfrac{[SO_3(g)]_{eqm}^2}{[SO_2(g)]_{eqm}^2 \times [O_2(g)]_{eqm}}$.
   b $mol^{-1}\,dm^3$.

8 a $\dfrac{[CH_3CHO(g)]_{eqm}^2}{[C_2H_2(g)]_{eqm}^2 \times [O_2(g)]_{eqm}}$.
   b $mol^{-1}\,dm^3$.
9 Temperature.

CONCEPT TEST

1 a $K_c = \dfrac{[CH_3COOCH_2CH_3][H_2O]}{[CH_3COOH][CH_3CH_2OH]}$

   $K_c = \dfrac{0.2 \times 0.2}{0.8 \times 0.8} = 0.0625$ (no units)

   (No units as vols cancel.)
   b Concentrated sulphuric acid absorbs water, so the amount of water will be decreased, so the equilibrium position will shift to produce more water and so increase the ester yield.

   c Catalysts do not change the equilibrium position so the yield is unchanged.
2 a The reaction is endothermic, so an increase in temperature shifts the equilibrium position to the right, so increasing the yield.
   b $K_c = \dfrac{[CO(g)] \times [H_2(g)]^3}{[CH_4(g)] \times [H_2O(g)]}$.
   c $mol^2\,dm^{-6}$.
   d An increase in temperature will increase the value of Kc as the forward reaction is endothermic.
3 a $K_c = \dfrac{[HI(g)]^2}{[H_2(g)] \times [I_2(g)]}$.
   b Doubling all the concentrations will have no effect as there the same number of molecules on both sides of the equation.

UNIT 9

RECALL TEST

1 a proton donor, a proton acceptor.
2 fully ionized.
3 a $pH = -\log_{10}[H^+(aq)]$.
   b $pK_a = -\log K_a$.
   c $[H^+] = 10^{-pH}$.
   d $K_a = 10^{-pK_a}$.
4 a pH = 1.3.
   b 1.3.
   c pH = 1.
   d 2.8.
5 $HCOOH(aq) \rightleftharpoons HCOO^-(aq) + H^+(aq)$.
6 pH = 12.7.
7 A buffer solution is a chemical mixture that resists changes in pH on addition of small amounts of acid or alkali. (Don't say it stops the pH changing, as the pH does change slightly).
8 pH = 4.76.
9 pH = 5.0.
10 See Fig 18.3 for answers.

CONCEPT TEST

1 a i proton acceptor.
      ii $pH = -\log[H^+(aq)]$, where $[H^+]$ is H^+ ion concentration in $mol\,dm^{-3}$.
      iii $K_w = [H^+_{(aq)}][OH^-_{(aq)}]$
   b pH = 1.7.
   c pH = 2.26.
   d pH = 6.76.
   e pH = 10.5.
2 a Acid added: $H^+(aq) + CO_3^{2-}(aq) \rightarrow HCO_3^-(aq)$.
      Alkali added: $HCO_3^-(aq) + OH^-(aq) \rightarrow CO_3^{2-}(aq) + H_2O(l)$.

b i $NaHCO_3 + NaOH \rightarrow Na_2CO_3 + H_2O$.
      The Na_2CO_3 dissolves to form CO_3^{2-} ions.
      $[CO_3^{2-}(aq)] = 0.05 \text{ mol dm}^{-3}$.
   ii The $NaHCO_3(aq)$ left over forms the HCO_3^- ions. So $[HCO_3^-] = 0.05 \text{ mol dm}^{-3}$.
   iii As the $[CO_3^{2-}(aq)] = [HCO_3^-]$, so pH = pK_a, so pH = 10.3.

UNIT 10

RECALL TEST

1 Lattice enthalpy is the energy change when one mole of an ionic solid is formed from its constituent gaseous ions.
2 The ions in CaO both have a 2+ charge so the lattice energy would be very much greater than that for NaCl, as in NaCl the ions only have a 1+ charge.
3 The Mg^{2+} is a much smaller ion than the Ca^{2+} ion. Mg^{2+} has one shell fewer than the Ca^{2+} ion.
4 The Born–Haber diagram would be similar to Fig. 12.1, but would require the production of only one gaseous Cl^- ion, so either one $\Delta H_a(Cl)$ or half the (Cl-Cl) bond energy.
5 Lattice energy is based on purely ionic compounds. Al_2Cl_6 is partially covalent.
6 The diagram should show that
   the sum of the hydration enthalpies
   = lattice enthalpy + solution enthalpy.
7 $Mg(OH)_2$ is insoluble because its ions are small so the lattice enthalpy is very large. $MgSO_4$ is soluble because the Mg^{2+} ion is small so the hydration enthalpy is large (while the sulphate anion is large so the lattice energy is low).
8 From MgO to BaO the lattice energy decreases as the cationic radius increases, making the cation less attractive.
9 A strong acid is an acid that is fully dissociated (ionised).
10 When strong acids and alkalis are dissolved in water they fully ionise, producing H^+ ions or OH^- ions. It is the reaction between these ions that produces heat when acids and alkalis are mixed.
11 All the acid reacts with NaOH. As ethanoic acid is a weak acid some molecules of the acid are not ionised. When the alkali is added the ethanoic acid molecules first ionise, which costs energy, so overall the reaction is less exothermic. $CH_3COOH \rightarrow CH_3COO^- + H^+$.

CONCEPT TEST

1 a The diagram should be similar to the $CaCl_2$ diagram (Fig. 12.1), except that the combined electron affinities are positive so the arrow should go up the diagram to the gaseous ions. CaO lattice enthalpy = $-3518 \text{ kJ mol}^{-1}$.

   b If you drew the diagram for $CaCl_3$ you would see that the 3rd ionisation energy was so large that it would make the ΔH formation massively endothermic. This means $CaCl_3$ would spontaneously turn into $CaCl_2$ and Cl_2.
   c Theoretical lattice energy is based on a purely ionic model. AgI is covalent so the experimental lattice enthalpy would be very different.
   d i $+1 \text{ kJ mol}^{-1}$.
      ii The energy released by the ions being surrounded by water supplies the energy required to separate the oppositely charged ions.
2 $+45.6 \text{ kJ mol}^{-1}$.

Unit 11

1 Entropy is a measure of disorder.
2 Entropy has the symbol S.
3 a increase
   b increase
   c increase
   d decrease
   e decrease
   f increase
4 **It must be negative.**
5 For an endothermic reaction to be feasible there must be an increase in disorder in the arrangement of the particles. It must that $T\Delta S^\ominus$ must be greater than $\Delta H^\ominus$. (Even though the enthalpy has a positive value).

6 a A more ordered arrangement of particles only occurs if the reaction is exothermic. . It must that $T\Delta S^\ominus$ must be greater than $\Delta H^\ominus$. Even though here $T\Delta S^\ominus$ is negative the enthalpy must be more negative.

7 When the ΔG indicates a reaction is feasible, but the reaction does not occur then the activation energy must be high, so the rate low.

CONCEPT TEST

1 a The reaction is spontaneous so ΔG must be negative.
   b i There is a decrease in the number of moles of gas so there is a decrease in entropy.
      ii $94 \text{ J mol}^{-1} \text{ K}^{-1}$.

2 a $\Delta G = \Delta H - T\Delta S$.
   b i $-26.2 \text{ kJ mol}^{-1}$.
      ii $14\,700 \text{ J mol}^{-1}$.
      iii The reaction is feasible at 1500 K as the ΔG is positive. It is not feasible at 500 K as the ΔG is negative.
      iv 1010 K (ΔG must be zero).

UNIT 12

RECALL TEST

1. The potential difference between a half cell and the standard hydrogen electrode under standard conditions when no current flows.
2. Same as Fig. 13.1.
3. Ag is positive electrode. Feasible reaction: Ni(s) + 2Ag$^+$(aq) → Ni^{2+}(aq) + 2Ag(s).
   e.m.f. = 0.33 volts.
4. a $H_2(g) + 2OH^-(aq) \rightarrow 2H_2O(l) + 2e^-$
   b $O_2(g) + 2H_2O(l) + 4e^- \rightarrow 4OH^-(aq)$
5. It must be due to the activation energy being high, so rate low.
6. If non-standard conditions are being used.
7. water, oxygen.
8. methane/hydrogen/a hydrocarbon, oxygen/air.
9. Fuel cells are much more energy efficient, more mobile and resist damage.
10. electrochemical series.
11. a Advantages: high efficiency, exhaust is pure water/little pollution/low carbon dioxide emissions. (also not dependent on unstable countries for energy supplies)
    b Disadvantages: (i) the hydrogen must be stored and transported safely, (ii) it may not be feasible to transport pressurised hydrogen liquid, (iii) the 'adsorber' or 'absorber' solids may not have a long enough lifetime; (iv) it costs to replace regularly the solids, (v) the disposal or recycling of the solids must be resolved; and the production costs may be too high, (vi) toxic chemicals may be required to produce the solids.
12. Political will on the part of governments. Social acceptance of fuel cells. Ignorance as some believe hydrogen is very much more dangerous then petrol.
13. Hydrogen storage: either it must be kept as high pressure liquid; adsorbed on the surface of a solid material; or adsorbed within a solid material. Alternatively the hydrogen is stored in a hydrogen rich fuel such as methanol, natural gas, or petrol, which are converted into hydrogen gas by an onboard 'reformer'.
14. Tidal, solar, nuclear or wind energy could produce electricity which then produces hydrogen.

CONCEPT TEST

1. a For oxygen to oxidise a halogen the oxygen $E^\ominus$ must be more positive than the halogen $E^\ominus$. The equation with the more negative $E^\ominus$ is where oxidation occurs. From the table only bromide and iodide are oxidised by oxygen under standard conditions.
   b The reaction may be feasible but the activation energy must be too high, so the rate is low.
2. a Iron(III) ions will oxidise the iodide ions, because the Fe^{3+}(aq)/Fe^{2+}(aq) $E^\ominus$ is more positive than the iodine/iodide $E^\ominus$. Oxidation occurs at the more negative $E^\ominus$.
   b The $E^\ominus$ indicate that this reaction does not occur under standard conditions, so non-standard conditions must be being used, e.g. the iodine and iron(II) ion concentrations may be more than 1 mol dm^{-3}.
   c The Fe^{3+}(aq)/Fe^{2+}(aq) $E^\ominus$ is more negative than F, G, H, so they will oxidise iron(II) ions to form iron(III) ions.
   d For oxygen to be reduced (gain electrons) the oxygen-containing $E^\ominus$ must be more positive than the iron-containing $E^\ominus$, so the iron-containing species that would be oxidised by oxygen are metallic iron, iron(II) ions, so the iron ends up oxidised to iron(III).
   e $2Fe(s) + O_2(g) + H_2O(l) \rightarrow 2Fe^{2+}(aq) + 4OH^-(aq)$
   f Zinc/zinc(II) ion $E^\ominus$ is more negative than iron $E^\ominus$ so the zinc will oxidise (corrode) in preference to the iron.
   g Nickel will protect the iron from water and oxygen which is necessary for rusting, until the nickel is scratched, because then the oxygen and water could make contact with the iron, starting rusting.
   h Disproportionation is the simultaneous oxidation and reduction of the same element, so the element must start at an intermediate oxidation number and then be oxidised and reduced. Fe +2 could be oxidised to Fe +3 and reduced to iron, but the $E^\ominus$ show that the opposite occurs. Fe +3 is another intermediate oxidation number, but the $E^\ominus$ indicate that Fe +3 would not disproportionate because the 2nd $E^\ominus$ is more negative than the 3rd.
3. a $Zn(s) + 2OH^-(aq) \rightarrow Zn(OH)_2(s) + 2e^-$
   b $MnO_2(s) + 2H_2O(l) + e^- \rightarrow Mn(OH)_3(s) + OH^-(aq)$
   c The zinc electrode.
   d The zinc electrode.
   e $Zn(s) + 2OH^-(aq) + 2MnO_2(s) + 4H_2O(l) \rightarrow Zn(OH)_2(s) + 2Mn(OH)_3(s) + 2OH^-(aq)$

UNIT 13

RECALL TEST

1. An element whose atom's last electron went into a d subshell is called a d-block element.
2. 4s.
3. a Cu: 1s^2 2s^2 2p^6 3s^2 3p^6 3d^{10} 4s^1.
   b Fe^{3+}: 1s^2 2s^2 2p^6 3s^2 3p^6 3d^5 4s^2.
4. A stable 3d^5 is formed when Fe^{2+} oxidises to Fe^{3+}, but it is difficult to oxidise Mn^{2+} to Mn^{3+} because it means losing an electron from the stable 3d^5.

5   a   octahedral,
    b   tetrahedral,
    a   square planar.
6   Element present, oxidation number, ligands present.
7   Complexes are formed when a central cation (or atom) accepts dative covalent (co-ordinate) bonds from ions (or molecules).
8   Ligands have lone pairs which form dative bonds with a central cation or atom in a complex.
9   A ligand that uses one lone pair per molecule in complexes.
10 Aqua H_2O, hydroxo OH^-, ammine NH_3, chloro Cl^-, oxo O^{2-}, cyano CN^-, thiocyano SCN^-, carbonyl CO.
11 a   optical,
    b   E-Z / geometric .
12 6.

CONCEPT TEST

1   a   A transition element is an element that forms at least one ion with a partially filled d subshell.
    b   Sc and Zn.
    c   transition metals form complexes; form coloured compounds; are used as catalysts / have a variable oxidation state.
    d   All forms. E-Z / geometric, and optical. (Also structural)
    e   Square planar complexes. (Actually octahedral also may show this). Not tetrahedral.
2   a   A ligand that uses one lone pair per molecule in complexes
    b   3d   4s
    Fe atom   (Ar)   [↑↓][↑][↑][↑][↑]   [↑↓]
    Fe^{2+} ion   (Ar)   [↑↓][↑][↑][↑][↑]   [ ]
    Fe^{3+} ion   (Ar)   [↑][↑][↑][↑][↑]   [ ]
    d   A stable $3d^5$ is formed when Fe^{2+} oxidises to Fe^{3+}, but it is difficult to oxidise Mn^{2+} to Mn^{3+} because it means breaking into the particularly stable $3d^5$.
3   a   *Either* diaminoethane, $NH_2CH_2CH_2NH_2$, *or* ethanedioate ions (oxalate ions) $C_2O_4^{2-}$.
    b   3.
    c   Optical isomerism.
    d   Octahedral.

UNIT 14

RECALL TEST

1   Fe^{2+} blue-green, acidified Fe^{3+} yellow, Co^{2+} pink, Cu^{2+} blue.
2   a   Redox.
    b   Deprotonation/ acid-base.
    c   Deprotonation/ acid-base.
    d   Ligand exchange/ ligand substitution.
3   Fe^{2+} YNYNY, Fe^{3+} YNYNN, Co^{2+} YNYYY, Cu^{2+} YNYYN. (Y = Yes, N = No.)

4   a $Fe^{2+}(aq) + 2OH^-(aq) \rightarrow Fe(OH)_2(s)$.
    b $[Cu(H_2O)_6]^{2+}(aq) + 4Cl^-(aq) \rightarrow [CuCl_4]^{2-}(aq) + 6H_2O(l)$.
    c $[Co(H_2O)_6]^{2+}(aq) + 4Cl^-(aq) \rightarrow [CoCl_4]^{2-}(aq) + 6H_2O(l)$.
    d $[Cu(H_2O)_6]^{2+}(aq) + 4NH_3(aq) \rightarrow [Cu(NH_3)_4(H_2O)_2]^{2+}(aq) + 6H_2O(l)$.
5   Ligand substitution is when one ligand replaces another ligand in a complex. The dative covalent bond of the original ligand breaks. The replacing ligand forms a dative covalent bond with the central cation of the complex.
6   $K_{stab} = \dfrac{[NiCl_4^{2-}]}{[Ni^{2+}][Cl^-]^4}$

    Units = $mol^{-4} dm^{12}$, or $dm^{12} mol^{-4}$

CONCEPT TEST

1   a   i $Fe(OH)_2$.
        ii Brown/red.
        iii $4Fe(OH)_2(s) + O_2(g) + 2H_2O(l) \rightarrow 4Fe(OH)_3(s)$.
    b   i $Fe^{2+} \rightarrow Fe^{3+} + e^-$.
        ii $MnO_4^-(aq) + 8H^+(aq) + 5e^- \rightarrow Mn^{2+}(aq) + 4H_2O(l)$.
        iii $MnO_4^-(aq) + 5Fe^{2+}(aq) + 8H^+(aq) \rightarrow Mn^{2+}(aq) + 5Fe^{3+}(aq) + 4H_2O(l)$.
2   a   Deprotonation/acid–base.
    b   Ligand exchange/ ligand substitution.
    c   By the addition of aqueous acid, e.g. dilute sulphuric acid.
    d   By the addition of conc. HCl or NaCl.
3   a   Zinc, iron, vanadium.
    b   acidified MnO_4^-.
    c   acidified MnO_4^- ; acidified $VO_2^+(aq)$.
    d   iron, zinc, $V^{3+}(aq)$, $V^{2+}(aq)$, V(s).
4   a $Cu^+(aq) + 4CN^-(aq) \rightleftharpoons [Cu(CN)_4]^{3-}$
    b   Units = $mol^{-2} dm^6$, or $dm^6 mol^{-2}$
    c $[Cu(CN)_4]^{3-}$ is the more stable.
    d $[Cu(CN)_4]^{3-}$.

UNIT 15

SYNOPTIC EXAM-STYLE QUESTIONS

1   a $Ag^+(aq) + Cl^-(aq) \rightarrow AgCl(s)$. (1)
    b   87.61, so X is strontium. (4)
    c   By adding chlorine water. This produces a brown iodine with iodide. To confirm the presence of iodine, *either* add starch which turn blue-black *or* add hexane, which produces a purple layer. (2)
    d   Lattice energy is for purely ionic compounds. Iron(III) chloride must have some covalent character. (2)

e CuCl$_4^-$. The different ligands change the energy-level gap within the 3d subshell. It is this energy-level gap which absorbs particular colours. If the gap changes then the colour absorbed changes, so the substance appears a different colour. (3)

f i By reduction and oxidation/electron transfer.
   ii By nucleophilic substitution.
   iii By nucleophilic substitution.
   iv By free-radical substitution.
   v By electrophilic substitution. (5)

g The hydroxide ions are nucleophiles (have a lone pair) so they will substitute the chloride, by nucleophilic substitution. The Cl on the benzene ring, Cl$_\text{C}$6$_\text{H}$4C$_\text{H3}$, is protected from the nucleophilic attack by the delocalised pi bonding which would repel the incoming lone pair. (3)

(Total 20 marks)

2 a −724 kJ mol^{-1}. (3)
  b i The equilibrium will shift to the left because the reaction is exothermic.
     ii The forward reaction rate will increase because increasing temperature increases the number of collisions with an energy greater than the activation energy.
     iii The reverse reaction rate increases for similar reasons to ii. (3)
  c The high temperature is required to produce an economic rate of production of HCN. The yield must be economic even though a lower temperature would increase the yield. (2)
  d The yield and rate must be high enough, so economic, at the stated pressure. Increasing the pressure would be expensive with little rate increase gained. An increase in pressure would lower the yield because there are more product gaseous molecules than reactants. (2)
  e i +1.
     ii Dicyano gold(I) ion.
     iii Covalent and dative covalent bonding.
     iv [Au(CN)$_2$]$^-$(aq) + Zn(s) → Au(s) + Zn^{2+}(aq) + 2CN$^-$(aq). (4)
  f When HCl(aq) and KOH(aq) react, only the H$^+$ and OH$^-$ react to form water because both compounds are fully ionised in water. As HCN is a weak acid it starts partially ionised. All the HCN reacts with the KOH so the unionised HCN ionises, which is an endothermic change, so the overall enthalpy is lower for the HCN with KOH. (2)

g i

$$[-CH_2=C-]_n$$
with COOCH$_3$ and CN substituents

(2)

   ii High temperature and high pressure, with an initiator. (2)

(Total 20 marks)

UNIT 16

RECALL TEST

1 a Accuracy is how close results are to the true value.
  b Precision is how many significant figures the data has.
  c Reliability is whether the data is consistent.

2 a QUAL: make sure you write down as many observations as possible; be ready to explain the observations.
  b QUANT: carry out practical work accurately; use the correct glassware; write down the units all the time; use the correct significant figures; measure temperature at the start and at the end; measure time at the start and at the end. Shake or stir mixtures.
  c EVAL: Link criticism to suggested improvements. Suggest the experiment is repeated, anomalies are ignore, an average taken, consistent results sought, larger masses or volumes used. There are many other possible answers. Shake or stir mixtures.

3 Consider the points below which are not exhaustive.
  a RATES: consider these points: measure time at the start and at the end; measure volumes accurately and consistently; control the temperature by using insulation, a lid, and a water bath; use the same surface area of solids. Shake or stir mixtures.
  b ENTHALPY: measure temperature at the start and at the end; control the temperature by using insulation, a lid, and a water bath; remember that a temperature rise indicates and an exothermic reaction which has a negative enthalpy.
  c TITRATION: titres are to two decimal place with the second figure being a 0 or 5; always add units; always wash burette and pipette with the solution you are to use; use the glassware correctly; use larger volumes or masses.
  d GAS VOLUME: ensure no gas escapes; collecting gas over water may be less accurate as the water may dissolve the gas or the temperature of the water may not be correct; ; use larger volumes or masses.

CONCEPT TEST

1. $0.256 \text{ mol dm}^{-3}$.
2. $I_2(\text{in KI(aq)}) + 2S_2O_3^{2-}(\text{aq}) \rightarrow 2I^-(\text{aq}) + S_4O_6^{2-}(\text{aq})$.
   Concentration iodine = $0.00538 \text{ mol dm}^{-3}$.
3. Concentration chloride ions = $0.0094 \text{ mol dm}^{-3}$.
4. 94.0% pure.
5. $0.0321 \text{ mol dm}^{-3}$.
6. **a** Percentage yield = 57%.
   b Dinitrobenzene was made, probably due to the temperature rising above the 60 °C required for nitration.
7. **a** Both the ethanoic acid and ethanol moles must be calculated. Ethanoic acid (using density) = 0.1748333 moles. Ethanol = 0.0.21739 mols. So the ethanol is in excess. The ethanoic acid limits the yield. Theoretical moles of ester = 0.1748333 moles. Theoretical mass of ester = 15.38533 grams. Yield = 15/15.38533 x 100 = 97.495450212% = 97.5% (to 3 significant figures).

   b i 0.1 / 10 x 100 = 1 %

   ii 0.01 / 10 x 100 = 0.1 %

   c 3 significant figures as the volume measured was to 3 figures.

   d (many other answers are possible)

Inaccuracy 1: volumes and masses were small.

Improvement 1: Use a larger volume and mass, say 50 cm³ and 50 grams.

Inaccuracy 2: No repeats.

Improvement 2: Repeat the experiment and take and average/ repeat until consistent/ repeat but ignore anomalies.

Inaccuracy 3: Glassware was inaccurate.

Improvement 3: Use a more accurate glassware. A pipette or burette

INDEX

ΔG, Gibbs free energy 62, 63, 64
ΔS, change in entropy 62, 68

absorbances in infrared spectra 13, 32, 33
accuracy 94 36
acid 4, 14, 15
acid–base reactions 14, 27, 50, 16, 22
activation energy 62, 69, 88, 51, 52, 34, 50, 51, 52, 58
addition 2, 8, 9 80
addition polymer 21, 20 27
airport security 33
alcohol 10, 15, 20
alcohol primary secondary tertiary 10 34
alkenes 2, 34
aluminium chloride 3
amide (link) 20, 21
amine 16, 22
amine primary secondary tertiary 16
amino acid 20, 22, 28
ammonia 9, 16, 46
analysis 50, 81
anhydride 1582
anionic 75
anionic radius 56, 58
antifebrin 19
aromatic 2, 3, 4
aspirin 15 14, 16
azo dye 434

base reactions 14, 27, 50
battery 63 51, 52
benzene 2, 3, 4 80
bidentate 75 14, 16
biodiesel 1534
Blood plasma 51
bond enthalpy 2, 56, 57
bond enthalpy 2, 60, 90, 58, 62
Born–Haber cycles 5664, 94
bromination 2, 3
bromine 2, 3, 4
Brønsted–Lowry theory 50 34
Buchner funnel 94
buffer 51, 52
buffer range 52

C-13 NMR 33
carbon monoxide 81
carbonate 4, 14, 51
carbonyl 8, 9, 10, 81

carbonyl compounds 8, 9, 10, 75
carboxylic acid 8, 14, 15
catalysts 3, 14, 45, 20, 22
cation 74
cationic radius 56, 58
character, covalent 57
chemcial tests 34
chemical shift 33
chiral 21, 22, 28
chloro 75
chromophore 4
cis (E-Z isomerism) 15, 27, 75
cis-platin 76 76
cobalt 74, 82
colorimeter 39
colour 4, 82
complex 74, 75, 76
compounds primary secondary tertiary 10
concentrated sulfuric acid 2, 3, 14
condenser 10, 93
conductivity meter 39
conjugate acid 50
conjugate base 50
co-ordinate bond 3
copper 68, 69, 74
coupling reaction 4 75, 82
coursework skills 93, 94, 95 86, 87
coursework tasks 93, 94, 95
covalent character 57
covalent character 57

d block 74, 75, 76
dative covalent (co-ordinate) bond 3, 74, 81
dative covalent bond 74
deuterium 33
diaminoethane 76
diazonium ion 4
dicarboxylic acid 20
dimer 14
dinitrophenylhydrazine (2,4-) 8
diol 20
disorder 62, 63, 64
distillation (heat under) 10
drugs 28
dry cells 63
dye 4
dynamic equilibrium 44

electrode potential 68, 69, 70
electron affinity 56
electronic configuration 74
enthalpy 2, 56, 57

enthalpy and experiments 94 58, 62
enthalpy change 56, 57, 58 63, 64, 94
enthalpy of atomisation 56 94
enthalpy of hydration 56
enthalpy of neutralisation 58
enthalpy of neutralisation 58
enthalpy of solution 56
entropy, S 62, 63, 64
environment 33
environmental analysis 33
enzyme 22, 28
equivalence point 52
esterification 4, 13, 14
esters 4, 13, 14, 15, 20
ethanal 8, 9, 10, 15, 20, 21
ethane-1,2-dioate (oxalate) ion 76 22, 28, 21
ethanoic acid 14, 50, 58
ethanoic acid 10, 15, 50
ethanoic anhydride 15 58
ethanol 8, 10, 94
ethene 8
ethylamine 16
evaluative tasks 93
experimental skills 93, 94, 95
E-Z isomerism (geometric) 15, 27, 75

fats and oils 15 76
feasible or spontaneous reactions 62, 68
filtering under reduced pressure 94
forensics 33
free radical substitution 3
fuel cell 64
gas chromatography, GC 32

GC, gas chromatography 32
Gibbs free energy, ΔG 62, 63, 64
gradient (rate) 39 68

[H] 111, 8, 10
haemoglobin 81 14, 26
half equations 68, 69, 70
half life 39
haloalkane
halogenation 3
heat under (reflux/ distillation) 10
hexadentate 75, 76
high pressure liquid chromatography, HPLC 32
high-resolution NMR 34

H-NMR (proton-NMR)   33
homolytic free radical
    substitution   3
HPLC, high pressure liquid
    chromatography   32
hydration   56, 58
hydration enthalpy   58
hydrogen bonding   8, 14, 21
hydrogen economy   64 28
hydrogen electrode   68
hydrogen fuel cell   64
hydrogencarbonate   51
hydrolysis   15

infrared absorption wave
    numbers   13, 32, 33
infrared spectroscopy   13, 32,
    33, 36
initial rate   40 36
ionic bond strength   56, 57,
    58
ionic character   56, 57, 58
ionic lattice   56, 57, 58
ionisation energy   56, 57, 58
iron   74, 80, 81
iron in haemoglobin   81 82
isoelectric point   22
isomerism E-Z isomerism   15,
    27, 75
isomerism geometric   15, 27,
    75, 76
isomerism structural   27, 28,
    32, 76

K_a 50
K_c 44
ketone   8, 9, 10
Kevlar   20 32, 33
Kind   5234
kinetics   38, 39, 40
kinetics of reactions   38, 39,
    40, 69, 93
K_{stab} 82 69, 93, 94
K_w 51, 94

lactic acid   20, 21, 22
Le Chatelier's principle   44, 69
ligand   74, 75, 80
low-resolution NMR   33, 81

Magnetic resonance imaging,
    MRI   34
manganese   73, 74
mass spectrometer   32
mass spectrum   32
mechanisms   2, 4, 8

medicine   28 9, 26
melting point apparatus   927,
    86
mobile phase   32
monodentate   75
monomer   20, 21
MRI, Magnetic resonance
    imaging   34

neutralisation, enthalpy of   58
nitrobenzene   4
nitrous acid   4
NMR   32, 33, 34
NMR, Nuclear Magnetic
    Resonance   33
Nuclear Magnetic Resonance,
    NMR   33
nucleophile   8, 9, 15
nucleophilic addition   9 16
nucleophilic substitution   16
Nylon   20

[O]   8, 10, 26
octadecenoic acid   15
olive oil   15
order   38
organic chemistry   2, 8, 14
organic mechanisms   2, 4, 8,
    20, 26
organic synthesis   26 9, 26
oxalate (ethane-1,2-
    dioate)ion   7627, 86
oxidation   68, 80
oxidation   8, 10, 26
oxidation number/state   68
    38, 68
oxidation variable   7469, 74,
    80, 82,
oxidising agent   68, 80

percentage error   93
percentage yield   93
pH   50
pH meter   39, 52
pharmaceuticals   28
phenol   4
phenylamine   4
phosphoric acid catalyst   26
phosphorus pentachloride   34
pi (π) bond   2
political obstacles   64
polydentate   75, 76
polymers   20, 21, 22
polymers addition   20, 25 28
polymers biodegradable   20
polypeptide   21, 22

potassium dichromate   8, 10,
    26
potassium permanganate   10
practicals   92, 93, 94
practicals techniques   92, 93,
    94
precipitate   80, 81
precipitation   80, 81
precision   93
propanone   8, 9, 10
protein   22
proton-NMR (H-NMR)   33

qualitative tasks   93
quantitative tasks   93

Radio waves   33
rate   38, 39, 40
rate constant   39 47, 62
rate determining step   3869
rate forward and backward   44
rate in practicals   93
reaction mechanism   2, 4, 8
reactions types (organic)   10
    9, 26
recrystallization   927, 38, 86
reducing agent   8, 10, 14
reduction   68, 80 68, 70
reduction   4, 8, 69
reflux (heat under)   10
reliability   94
repeating unit   20, 21, 22
retention time   32
reversible
reversible reactions   44, 50
R_f value   32

safety   94
saponification   15
saturated fats   15
side effects   28
silver nitrate   8, 9, 34
soap   15
social obstacles   64
solution enthalpy   58
solvent front   32
space probes   33
spectroscopy   32, 33, 34
spontaneous or feasible
    reactions   62, 68
stability   62
standard electrode
    potential   62
stationary phase   32
strong acid   50
strong base   50
sulfuric acid   2, 3, 14

sulfuric concentrated   2, 3, 14
supercritical carbon dioxide   28
surrounds   63
synoptic   87, 88, 89
synthesis   26

TBP (2,4,6-tribromophenol)   4
Terylene   20
test for acid functional group (-COOH)   34
test for alcohol (O-H)   34
test for alkenes   34
test for haloalkane   34
Tetramethylsilane, TMS   33
Thin-layer chromatography, TLC   32

titration calculations   89, 92, 96
TLC, Thin-layer chromatography   32
TMS, Tetramethylsilane   33
trans   15, 27, 75
transition elements   74, 75, 76, 76
transition metal   8, 74, 75, 80, 81
trans-platin   76 76, 80, 82
tribromophenol (TBP)   481, 82, 92
triglyceride esters   15

unifying concepts   87, 88, 89

vegetable oil   15, 93

water-soluble organic molecules   28
wavenumber   37
weak acid   50
weak base   50

yeast   28
zinc   68, 74
zwitterion   22

The Periodic Table of the Elements

Key

| Molar mass g mol⁻¹ |
| **Symbol** |
| Name |
| Atomic number |

Period	Group 1	Group 2												Group 3	Group 4	Group 5	Group 6	Group 7	Group 8
1	1 **H** Hydrogen 1																		4 **He** Helium 2
2	7 **Li** Lithium 3	9 **Be** Beryllium 4												11 **B** Boron 5	12 **C** Carbon 6	14 **N** Nitrogen 7	16 **O** Oxygen 8	19 **F** Fluorine 9	20 **Ne** Neon 10
3	23 **Na** Sodium 11	24 **Mg** Magnesium 12												27 **Al** Aluminium 13	28 **Si** Silicon 14	31 **P** Phosphorus 15	32 **S** Sulphur 16	35.5 **Cl** Chlorine 17	40 **Ar** Argon 18
4	39 **K** Potassium 19	40 **Ca** Calcium 20	45 **Sc** Scandium 21	48 **Ti** Titanium 22	51 **V** Vanadium 23	52 **Cr** Chromium 24	55 **Mn** Manganese 25	56 **Fe** Iron 26	59 **Co** Cobalt 27	59 **Ni** Nickel 28	63.5 **Cu** Copper 29	65.4 **Zn** Zinc 30	70 **Ga** Gallium 31	73 **Ge** Germanium 32	75 **As** Arsenic 33	79 **Se** Selenium 34	80 **Br** Bromine 35	84 **Kr** Krypton 36
5	85 **Rb** Rubidium 37	88 **Sr** Strontium 38	89 **Y** Yttrium 39	91 **Zr** Zirconium 40	93 **Nb** Niobium 41	96 **Mo** Molybdenum 42	99 **Tc** Technetium 43	101 **Ru** Ruthenium 44	103 **Rh** Rhodium 45	106 **Pd** Palladium 46	108 **Ag** Silver 47	112 **Cd** Cadmium 48	115 **In** Indium 49	119 **Sn** Tin 50	122 **Sb** Antimony 51	128 **Te** Tellurium 52	127 **I** Iodine 53	131 **Xe** Xenon 54
6	133 **Cs** Caesium 55	137 **Ba** Barium 56	139 **La** Lanthanum 57	178 **Hf** Hafnium 72	181 **Ta** Tantalum 73	184 **W** Tungsten 74	186 **Re** Rhenium 75	190 **Os** Osmium 76	192 **Ir** Iridium 77	195 **Pt** Platinum 78	197 **Au** Gold 79	201 **Hg** Mercury 80	204 **Tl** Thallium 81	207 **Pb** Lead 82	209 **Bi** Bismuth 83	210 **Po** Polonium 84	210 **At** Astatine 85	222 **Rn** Radon 86
7	223 **Fr** Francium 87	226 **Ra** Radium 88	227 **Ac** Actinium 89															

140 **Ce** Cerium 58	141 **Pr** Praseodymium 59	144 **Nd** Neodymium 60	(147) **Pm** Promethium 61	150 **Sm** Samarium 62	152 **Eu** Europium 63	157 **Gd** Gadolinium 64	159 **Tb** Terbium 65	163 **Dy** Dysprosium 66	165 **Ho** Holmium 67	167 **Er** Erbium 68	169 **Tm** Thulium 69	173 **Yb** Ytterbium 70	175 **Lu** Lutetium 71
232 **Th** Thorium 90	(231) **Pa** Protactinium 91	238 **U** Uranium 92	(237) **Np** Neptunium 93	(242) **Pu** Plutonium 94	(243) **Am** Americium 95	(247) **Cm** Curium 96	(245) **Bk** Berkelium 97	(251) **Cf** Californium 98	(254) **Es** Einsteinium 99	(253) **Fm** Fermium 100	(256) **Md** Mendelevium 101	(254) **No** Nobelium 102	(257) **Lr** Lawrencium 103

www.ingramcontent.com/pod-product-compliance
Lightning Source LLC
Chambersburg PA
CBHW081418300426
44109CB00019BA/2343